Waste Management

Edited by
Er Sunil Kumar

Waste Management

Edited by Er Sunil Kumar

Republished by InTech

Janeza Trdine 9, 51000 Rijeka, Croatia

Technical Editor Teodora Smiljanic

Cover Designer Dino Smrekar

Additional hard copies can be obtained from orders@intechopen.com

Waste Management, Edited by Er Sunil Kumar
p. cm.
ISBN 978-953-7619-84-8

For EU product safety concerns:
IN TECH d.o.o., Prolaz Marije Krucifikse Kozulić 3, 51000 Rijeka, Croatia,
info@intechopen.com or visit our website at intechopen.com.

Meet the editor

Er. Sunil Kumar is an expert in Environmental Engineering and Management with specialization in Solid and Hazardous Waste Management. He is a member of Editorial Boards of several international journals, such as Environmental Monitoring and Assessment, The Netherlands, and Open Waste Management Journal, UK. He is the Regional Editor for Asian countries in the International Journal of Process Wastes Treatment, Serial Publications, New Delhi, and is an expert member in Development of a Fact Base of Greenhouse Gas Emissions from Waste Sector in India leading up to 2030. He has served as Editor for several international publications, among which the Special Issue for Environmental Monitoring and Assessment, The Netherlands; Journal of Air and Waste Management Association, USA and International Journal of Environment and Pollution, and others.

Preface

Solid Waste Management is one of the essential obligatory functions of the Urban Local Bodies/Municipal Corporation. This service is falling too short of the desired level of efficiency and satisfaction resulting in problems of health, sanitation and environmental degradation. Due to lack of serious efforts by town/city authorities, garbage and its management has become a tenacious problem. Moreover, unsafe disposal of garbage and wastewater, coupled with poor hygiene, is creating opportunities for transmission of diseases. Solutions to problems of waste management are available. However, a general lack of awareness of the impact of unattended waste on people's health and lives, and the widespread perception that the solutions are not affordable have made communities and local authorities apathetic towards the problems.

The aim of this Book is to bring together experiences reported from different geographical regions and local contexts. It consolidates the experiences of the experts from different geographical locations *viz.*, Japan, Portugal, Columbia, Greece, India, Brazil, Chile, Australia and others.

It is hoped that this publication will open the eyes of the citizens in this part of the world to the increasing menace posed by the lack of waste management systems and inspire them to do their share to make such systems operative. It also hopes to instill a moral compulsion among policy makers to make waste management a part of their government's development policy.

Editor

Sunil Kumar
Scientist, National Environmental Engineering Research Institute (NEERI)
[Council of Scientific and Industrial Research (CSIR)]
Kolkata Zonal Laboratory
I-8, Sector "C", East Kolkata, EM Bypass
Kolkata , 700 107, India
E-mail: s_kumar@neeri.res.in
sunil_neeri@yahoo.co.in

Biographical notes: Sunil Kumar is a Civil and Environmental Engineer with 12 years of professional experience in municipal solid waste management. He engages in environmental impact assessment and management, design of sanitation system including solid and hazardous waste management.

Contents

1

The Threshold Target Approach to Waste Management in Emerging Economies: Pragmatic, Realistic, Appropriate

Manfred Fehr
Federal University at Uberlândia
Brazil

1. Introduction

Municipal solid waste is a commodity resulting from human endeavor and will always exist in the world's cities. The rate of production is commonly reported on a per-person-per-day basis, which implies that it will grow together with the population. In the not too distant past, waste was conceived of as a nuisance that could be removed by collection followed by tipping at a site not visible to the population: a landfill at best. This concept of a landfill as waste receiver has been challenged in recent decades for reasons of space availability and of collateral effects on soil, water and air. World summits have addressed the topic of waste within the general subject of sanitation, and have set targets for the gradual expansion of collection services to urban residents. Specifically, the directives resulting from the World Summit on Sustainable Development (WSSD 2002) asked to halve by the year 2015 the proportion of people not served by sanitation. Global intellectual movements such as the Zero Waste International Alliance (ZWIA 2009) have gone much further in their quest for sustainable waste management. In fact, several cities have adhered to those movements by passing local legislation that requires gradual reduction of waste disposal until reaching a zero waste situation within a timeframe in the order of fifteen years. Against this background of increasing demand for results, city administrators face enormous management challenges.

Although waste is produced and handled everywhere in the world, different countries are moving at different speeds to set and meet waste management targets. Southern countries in general are at the stage of moving from dumpsites to landfills, whereas Northern countries are already phasing out the landfills by efficient recycling and mechanical-biological pretreatment procedures. As a consequence, the management challenges differ, and appropriate solutions need to be developed for different groups of countries. Waste collection service is not yet universal in Southern countries. Consequently, expanding it as required by summit directives fatally implies more landfill space. This underlines the argument of different speeds and different directions. Landfills are still increasing in the South, whereas they are already decreasing in the North. Different countries at present are on different sides of the turning point, which represents complete collection service. Along comes the zero waste movement and sets the pace with pioneering cities towards eliminating waste altogether. The basic argument of the present study is that the move sought by the zero waste movement is only realistic when it starts beyond the turning point, i.e. in Northern countries. Very few cities of the developing world are prepared at this time to make such a move, almost a utopia to them. They need a viable alternative to zero waste, which will allow them to set intermediate targets on the long journey to sustainable waste management. The present study is designed to provide those intermediate targets.

Waste management research by this author in Brazil has identified thresholds for different components of municipal waste that are suitable for landfill diversion targeting. What is a threshold in this context? It represents the percentage of each waste component that can be moved through the reverse logistics chain by market forces without the necessity of public funds. This is a new concept in waste management. Once the threshold is known, it becomes the natural target for landfill diversion within a timeframe to be established by the municipal administration. The quantity of the waste situated beyond the threshold is left as a future challenge to be tackled in due time. The utopia has been eliminated, and the financial bottleneck of the city budget has been bypassed. By way of examples, the research has identified the threshold for domestic waste as 67% and that for construction debris as 90% of collected quantity. The methods used for determining the thresholds are described, and administrative procedures for reaching the threshold targets are outlined. The procedures include, but are not restricted to, the following activities: Establish the threshold for each waste component and create local legislation to enforce the corresponding separation at the source. Create incentives for reverse logistics operators to absorb the material made available by source-separation with possible use of funds liberated by reduced landfilling. Design landfill capacity only for waste produced beyond the threshold. Put municipal waste management activities into the hands of marketing and accounting professionals.

Even without the pretension to reach zero waste, the challenges for cities of emerging economies are enormous. This study is a modest contribution to meeting the challenges.

The success or failure of urban waste management may be viewed as an indicator of sustainability.

Indicators have meaning if they incorporate a mixture of physical, economic and social data that evaluate changes in time and promote actions. Many measurable indicators for the state of development of a nation or country have been proposed and applied. Two of them are officially used by the United Nations to classify countries. They are the Gross National Product and the Human Development Index. The former only has meaning when applied to a nation, and even so has its shortcomings. The industrial and service outputs are not necessarily sustainability indicators. Large quantities of throw-away products that increase required landfill space contribute to the gross national product, but are not representative of a sustainable society (Kanitz 2006).

The Human Development Index, apart from being applied to countries, is also used to classify cities within a country. It measures the life expectancy, the frequency of school attendance, the degree of literacy, the infant mortality rate and the average individual income. All of them are considered indicative of the general state of living standard, and are perfectly appropriate for a municipal context in terms of determining the degree of sustainability.

The measure of relative quality of life is included in the Mercer Index that classifies the World's cities according to the parameters of security, health services, basic sanitation, air pollution, education and transport facilities (Report 2004).

Perhaps the most widely known sustainability indicator at present is the Ecological Footprint that measures the energy and material consumption of a society in terms of land area required to satisfy the demand (Rees 1996).

In 2004, this author's team proposed a specific set of indicators for urban sustainability that may be determined by local diagnosis and continuous data collection (Fehr et al 2004). As they are meant to identify absolutely defined situations, they may also be referred to as identifiers. Table 1 is reproduced from that paper to illustrate the degree of quantification aimed at.
The present study describes research carried out to more closely qualify the third identifier in Table 1: Landfill diversion of solid waste is in excess of 70%. After evaluating the present solid waste situation in a city, the study pursued the objective to define a threshold value of diversion, which separates results possible with private initiatives from results achievable only with public intervention.

Population growth is under control.
The public transportation system is of high quality.
Landfill diversion of solid waste is in excess of 70%.
All liquid effluents are treated.
Air quality is monitored.
Fresh water demand is monitored and controlled.
The public education system has high student and teacher satisfaction
Public health care is accessible and of high quality.
Citizens are socially and politically active.
Energy supply and demand are monitored and controlled.
Public recreation areas are available in all sectors of town.
Rivers and creeks are under official protection.

Table 1. Basic municipal sustainability identifiers (Fehr et al 2004)

2. Materials and methods

It is acknowledged that the landfill is part of the ecological footprint in as much as it sets aside land area for depositing the city's refuse. The rate of deposit is appropriate to be used as a sustainability indicator, since it evaluates changes in time and promotes actions. To illustrate the change in time, the land distribution of the Planet may be invoked. In the year 2000, seven percent of the Earth's surface was deemed useful for agriculture, industrial activities and city dwelling, for a total of $36*10^6$ km^2, and there were $6*10^9$ people (Fehr 2003). The urban area proper occupied 5% of the arable land, or $1.8*10^6$ km^2, and the landfills, in turn, occupied 1% of urban area or $1.8*10^4$ km^2. The World's population is expected to level off at $10*10^9$ people around 2080 (Doyle 1997).
At this rate of population growth and trash production, a typical city will have to construct a new landfill every 16 years. The land occupied by landfills will thus reach 6% of urban area by 2080. The change in time is impressive. The urban terrain per person in 2000 was $1.8*10^6 / 6*10^9 = 300$ m^2. In 2080 the urban terrain per person will be $1.8*10^6 / 10*10^9 = 180$ m^2, of which landfills will occupy six percent or 11 m^2. This is the fatality resulting from constant waste management policies in the 80-year period. The urban area available for living will shrink from 300 - 3 = 297m^2 to 180 - 11 = 169 m^2 per person. As this sequence of events is a road sign to collapse, the landfill area, or alternatively the rate of landfill diversion of waste, is a strong candidate for sustainability indicator. How does this indicator promote action? With proactive waste management models, about 80% of urban waste may

today and in the near future be diverted from the landfills (Fehr and Calçado 2001). This means that landfill area would be maintained at $11*(1-0.8) = 2\ m^2$ or approximately one percent of urban terrain per person from 2000 to 2080. The indicator will allow for monitoring the progress and the success of the proactive waste management model.

The remainder of this chapter is dedicated to describing experiments carried out with the objective to build at least one possible version of the management model required. The relativity of sustainability comes into play here. If sustainability were considered a state function or a fixed condition, then a city could only be sustainable or unsustainable. If sustainability were considered a process, then intermediate or fractional approaches to sustainability could exist. In the case made here, if the city in question was sustainable in 2000 with one percent of its space dedicated to landfills, and the management model was applied successfully, the city would still be sustainable in 2080 with the same one percent of its space dedicated to landfills, other factors notwithstanding.

As the implementation of the model takes up time, the required landfill space will exceed one percent of urban area for a number of years and finally return to one percent and level off. The function of the indicator is to register this change in time in order to promote the corresponding actions. What is to be measured is the evolution of the fraction of urban waste diverted from the landfill and recycled as a function of time, and this fraction will be representative of the approach to sustainability related to urban waste, with 0.67 being a satisfactory starting target for domestic waste as will be explained shortly.

The threshold for landfill diversion was arrived at through research on urban waste composition for raw waste and for sorted waste. Tables were developed to show the evolution of waste production, waste collection and landfill diversion in the city with and without the use of threshold targeting. Threshold targeting is presented as a pragmatic tool for the development of waste management schemes.

3. Results and discussion

3.1 Household waste

The experiments carried out with household waste management identified a threshold for landfill diversion at 67% of total waste produced. The meaning of the threshold is as follows. The indicated diversion of 67% represents the maximum rate achievable with strictly private initiatives, or bottom-up management procedures, which may be either spontaneous or stimulated. In order to move above the threshold, intervention by the city administration is required. This leads directly to proposals for the diversion of the remaining 33% of city trash.

Existing management models for household waste were analyzed, but none evidenced a landfill diversion potential above the present value of 15% for selective collection of inert material in Brazil. A sharp paradigm shift was required to raise this potential to values above 70% and thus create the prospect of sustainable situations. The research led to the *Divided Waste Processing* (DWP) model as means to meet the challenge (Fehr & Calçado 2001). This model differentiates between humid and dry material in the waste stream, or in biological terms between biodegradable and inert material. Thus the model requires the use of only two recipients, one for each portion, and the collection and processing operations maintain them separated all the way to their respective destinations. Once the management model had been elaborated, tests of its functionality were initiated with the objective to demonstrate the landfill reduction achievable. Destinations of source-separated material were animal feed and compost for the humid part, and informal reverse logistics for the dry part.

As the key to success is the correct source separation, the challenge was clearly educational and was faced and met as such. The research started in apartment buildings, was then extrapolated to a street, and recently arrived at the stage of using schools as multipliers of the model. In all those communities the model confirmed its consistency as it pointed to the same theoretical diversion potential of 83% even if to date this level has not been reached. This communication relates the experience gained, the arguments used with the communities and the results obtained with the active environmental education procedure. It opens up the prospect to amplify the application of the model to other sectors and eventually to the whole city.

The experiment described here is original in the sense that it is an entirely private initiative that takes the message of its results from the bottom upwards into the municipal administrative hierarchy. Traditional models follow the inverse direction.

In the first test community, the divided waste-processing model was functional in 60 apartments after 4 person-months of dedication. The humid and dry fractions of raw waste stabilized at 68% and 32% respectively. The behavior change was obtained and perpetuated by the building administrators through the constant communication of results. The effective participation rate of residents was above 80%. This level was considered excellent in a context where a completely new model with voluntary participation was applied. The behavior model of the participating families at present is as follows. In the apartments, all waste is rigorously separated into humid and dry parts. Each family uses a pail furnished by the administration to collect biodegradable waste for a day. At predetermined hours, the employees collect this pail. All inert waste is left at the collection point on each floor at any time of day and in any form of packaging. The employees transfer the contents of the pails to a barrel, which is taken away daily by a farmer who uses the material as animal feed. All inert material from the floors is transferred to a collection cart, which then is left at the disposal of selected waste retailers who take away approximately half of this material for their recycling businesses. What remains each day represents approximately 40% of all waste and is left at the curbside for official collection by the municipal vehicles that take it to the landfill.

In the second test community, the main factor of success was the insistence of the research team with the necessity of separation prior to collection. It happened that some residents did not separate their material for the programmed collection, and when questioned responded that the team had not visited them or left a message the day before. This example illustrates very well the difficulty of changing established thinking models.

In the work with the school communities, compost was prepared in the school yards from the source-separated biodegradable waste. The model turned out to be a powerful learning tool. All participating students now leave primary school with the baggage of practical experience of fabricating a useful product and with the conviction of having contributed their share to the reduction of the landfill. The compost is available for gardening in the school and in the neighborhood, and is introduced to the community as a product of what only a short time ago they used to call garbage.

In the case of domestic waste, the composition report is critical to setting diversion targets or thresholds. This is so because the composition of the waste depends on the amount of human intervention in its evolution. This research experimented with several stages of waste sorting and the corresponding compositions with the results reported in the following section.

Raw waste is what families discard in their waste baskets and leave for collection without any sorting effort. The analysis of this raw waste may yield three different types of composition reports. The first type of report results from the separation of raw waste into biodegradable material and biologically inert material. In the first test community this report produced the numbers shown in Table 2. The second type of report results from the separation of raw waste by substance. The same waste from the first community was analyzed for contained substances and produced the numbers shown in Table 3. The third type of report results from the separation of raw waste by utility. The same waste from the first community was analyzed for utility and produced the numbers shown in Table 4.

material	weight percent
biodegradable matter	68
biologically inert material	32

Table 2. Composition report by biodegradability for raw domestic waste

The evolution of the significance for decision making of the successive reports is apparent. The numbers in Tables 2 and 3 are basically of academic nature. They do not support waste management decisions on landfill diversion potential or educational efforts of source-separation. Table 4, to the contrary, supports such decisions. The educational effort required to reduce food waste by consumers is hidden in the lost food item. The landfill diversion potential may be read off Table 4 as 58+10+15=83%. This information is of utmost importance to the construction of a management model. It tells the administration that with an adequate model only 17% of present landfill capacity will be required in the long run.

material	weight percent
biodegradable matter	68
plastics	10
paper and cardboard	9
glass	4
textiles	3
metals	2
miscellaneous	4

Table 3. Composition report by substance for raw domestic waste

material	weight percent
food scraps to composting	58
lost food to further use	10
used packaging to recycle	15
trash to landfill	17

Table 4. Composition report by utility for raw domestic waste

There is however one basic shortcoming to all these reports, which is not visible to the unsuspecting observer but was identified by this research. All foregoing reports refer to raw waste. Consequently, no information is available on the success or failure of an effort to really separate the waste at the source into the categories listed. The indications derived for the municipal waste management model remain hypothetical. In order to advance, this research experimented with source-separation in the first test community for several months. Families were instructed to separate their waste into biodegradable matter and inert matter and deliver the two parts to the building administration for recycling. Containers were provided for the two types of waste, and the building employees collected them daily for screening. Instructions to various families were repeated to ascertain the procedure. After four months of experimentation, the following conclusions became available. It is impossible to obtain the collaboration of all families. An adhesion of 80% has to be considered excellent. The separation procedure at the source, even with the simple request of only two recipients, presents a heavy intellectual burden to most apartment dwellers. They do make an effort, but the success is only partial. Several items of waste end up in the wrong container. The building employees have to screen the delivered material and proceed with an additional separation before handing the sorted material over to reverse logistics operators. The result of this experiment was the two-step sorted-waste composition report shown in Table 5. In contrast to the previous reports, this one can be considered a management tool. It defines the landfill diversion potential as 67% of domestic waste, arrived at by experimental source-separation, and therefore this is a reliable number. The best raw waste composition report cited this potential as 83%, which would lead to erroneous decisions by waste managers. The trash item in Table 5 refers to material not separated at the source in spite of correct instructions and goodwill, as well as to items that are not recyclable at this time.

material	weight percent
biodegradable matter for composting	47
recyclable matter for reverse logistics	20
trash temporarily for landfilling	33

Table 5. Composition report for source-separated domestic waste

The key words in Table 5 are *source-separated* and *temporarily*. The former means that this composition represents the best separation result possible at this time in households, independently of the raw waste composition prevailing in the city. The latter means that the trash is not necessarily improper for recycling. It simply is not being separated at this time, thus opening targeting options to the municipal administration. Table 5 tells the administration that 67% of domestic waste can be recycled by private initiatives if the pertinent incentives were created. The remaining 33% represent the target for official intervention. Several options are available for diverting this material from the landfill. Some are obvious, as e.g. educational efforts to improve source separation, and policy tools to make more trash items attractive to reverse logistics. In the worst case, landfill capacity has to be provided to tip the 33% trash. This is the threshold mentioned at the beginning, and is the original contribution of this research to the science of waste management. As an indicator, the threshold of 67% landfill diversion, if reached, tells that the city administration has stimulated the private sector to contribute its maximum expected share to domestic waste recycling.

The expected waste movement at the stage where the threshold is reached is depicted on Figure 1. How can this stage be reached in practice? The proposal resulting from this research makes use of timeframes and annual targets that will lead to the threshold situation within those timeframes. The threshold target represents an intermediate stage between bulk tipping and "zero waste" offered to municipal administrations as a realistic and appropriate alternative to world summit directives. In order to illustrate the concept of timeframes and annual progress requirements, Tables 6, 7 and 8 will be explained now.

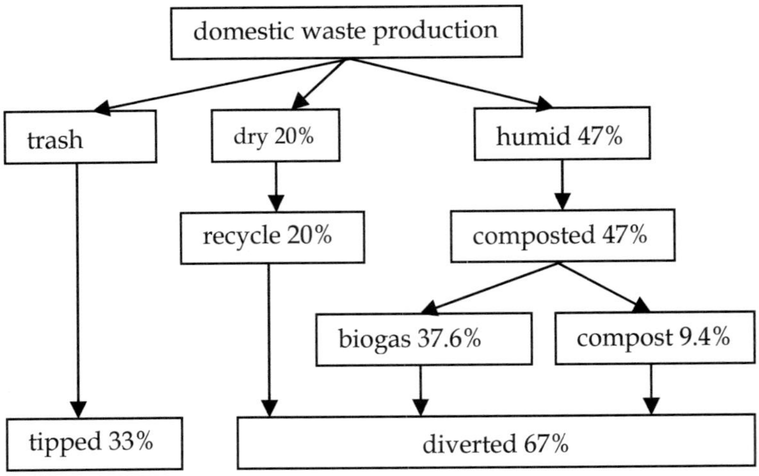

Fig. 1. Material movement of domestic waste at threshold situation

sequence of years	waste production preview kt/year	waste collection target according to WSSD directive kt/year	waste dumping target according to WSSD directive kt/year
0	182.5	146.0	36.5
1	183.8	148.4	35.4
2	185.1	150.8	34.3
3	186.4	153.2	33.2
4	187.7	155.7	32.0
5	189.0	158.2	30.8
6	190.3	160.7	29.6
7	191.6	163.3	28.3
8	193.0	166.0	27.0
9	194.3	168.7	25.6
10	195.7	171.4	24.3
11	197.1	174.2	22.9
12	198.5	177.0	21.5
13	199.8	179.8	20.0
14	201.2	182.7	18.5
15	202.6	185.7	16.9
16	204.0	188.7	15.3
17	205.5	191.8	13.7
18	206.9	194.9	12.1
19	208.4	198.0	10.4
20	209.8	201.2	8.6
21	211.3	204.5	6.8
22	212.8	207.8	5.0
23	214.3	211.1	3.1
24	215.8	214.5	1.2
25	218.0	218.0	0

Table 6. Production, collection and dumping targets from WSSD directives

Table 6 shows the base case of a municipality that adhered to the WSSD directive to halve, by 2015 (year 13) the proportion of residences not served by collection. In year 0 there were 500,000 inhabitants producing 1.0 kg per person per day of waste or 182.5 kilo tons (kt) per year. The collection service was available to 80% of the residences, which means that 182.5*0.8=146.0 kt were collected per year, and the remaining 20% or 36.5 kt per year were dumped by residents at unauthorized locations. The population and with it the waste production were increasing by 0.7% per year. The target set by the world summit required to reach 90% collection service by year 13. The necessary yearly collection expansion was found from equation 1

$$0.8 * x^{13} = 0.9 \qquad x=1.0091 \tag{1}$$

Considering the waste production increase of 1.007 per year, the resulting collection effort was defined as an annual increase of 1.0091*1.007.

As an example, for year 1 the production was 182.5*1.007=183.8 and the collection had to reach 146.0*1.007*1.0091=148.4 kt. The collection target for year 13 was set as $146.0*1.007^{13}*1.0091^{13}$ or 179.8 kt. This represents the required 90% of the 199.8 kt produced, with the remaining 10% or 20.0 kt being dumped.

Although the summit directive was satisfied, the exercise was extended at the same collection expansion until complete collection service would be reached. This occurred in year 25 (2027), when all waste produced would be collected and no more dumping would exist. This base case confirmed the following facts. In the city under study, it would take 25 years to reach complete collection if the service expansion required by the directive were extrapolated beyond year 13. In this model, all collected material is tipped at the landfill.

Table 7 shows the concept of threshold targeting. Here the collected material is partly tipped and partly diverted from the landfill. The city administration may choose any timeframe it deems reasonable to reach the target. In the example of Table 7 the timeframe was taken as

sequence of years	collection target from WSSD directive kt/year	threshold target for tipping kt/year	threshold target for diversion kt/year	progress of diversion %
0	146.0	146.0	0	0
1	148.4	140.8	7.6	5.1
2	150.8	135.3	15.5	10.3
3	153.2	129.5	23.7	15.4
4	155.7	123.7	32.0	20.6
5	158.2	117.4	40.8	25.8
6	160.7	111.0	49.7	30.9
7	163.3	104.4	58.9	36.1
8	166.0	97.6	68.4	41.2
9	168.7	90.4	78.3	46.4
10	171.4	83.1	88.3	51.5
11	174.2	75.4	98.8	56.7
12	177.0	67.5	109.5	61.9
13	179.8	59.3	120.5	67.0

Table 7. Tipping and diversion targets to reach threshold in 13 years

13 years. The first column repeats the waste collection requirement from Table 6. The next two columns illustrate how collected waste divides into tipping and landfill diversion.

The threshold to be reached in year 13 is the experimental value of 33% tipping and 67% diversion. Upon dividing 67% by 13 years, the annual diversion expansion is found to be 5.153846%. Excess significant digits are carried here to reduce round-off errors. As an example, the required diversion for year 1 is 148.4*0.05153846=7.6 kt, and the tipping rate has to be 148.4*(1-0.05153846)=140.8 kt. By the time year 13 is reached, the tipping rate would have been reduced to 179.8*(1-0.05153846*13)=59.3 kt or 33% of collected waste. The table shows the evolution of the diversion effort over the years and allows for precise targeting to reach the threshold within the chosen timeframe. No utopia is involved. The threshold value is experimental, and the required actions can be correctly dimensioned.

What are these actions? This research has contributed the most important one, namely promote source separation. The experimental work with the test communities showed that source separation drives reverse logistics, and not vice versa. The logistics for inert material is already in place and is completely private. It recycles all inert material conveniently separated at the source. The action of the public administration will be restricted to promoting source separation, either by local legislation or by educational campaigns. The model allows for ample time. Table 7 shows that 20% diversion is to be reached in year 4. This corresponds to the amount of inert material in the waste according to Table 5. Consequently, four years are available to divert all inert waste from the landfill by fostering source separation and stimulating reverse logistics: a reasonable task.

The logistics for biodegradable material is still incipient. It has to be developed within the remaining nine years. As the progress indicated in Table 7 is gradual, nine years are a reasonable period. Apart from stimulating more farmers to take away the biodegradable waste from restaurants and other commercial establishments, it will be necessary to support private composting facilities. There is no point in arguing that these facilities are not by themselves economically feasible. What has to be done is to dimension the savings resulting from reduced landfill construction and maintenance, and apply them to buying compost from the private operator at a price that will keep him or her in business. Other schemes may be invented. All depends on the creativity of the responsible administrators. What the threshold approach has provided is the realistic target for diversion. It has taken the guesswork out of waste management. Table 7 also implies result reporting. This is an essential part of modern waste management. It holds the public administration responsible for reaching the annual targets.

Another illustration of the use of this concept is presented in Table 8. It relates to a municipality that started to adhere to the summit directive in year 0 and progressed as planned until year 8, just as illustrated in Table 6. No diversion was considered until then. Starting in year 9, the administration decided to initiate threshold targeting. The tipping rate had been increasing as expected, but the outlook beyond year 8 called for gradual reduction. The timeframe chosen was 17 years until year 25, which was identified in Table 6 as the turning point where collection service would be complete. Following the threshold concept, from year 9 onward, the collected portion of waste was to be partly tipped and partly diverted, with the tipping rate reaching 33% by the end of the period.

The arithmetic for Table 8 is as follows. By dividing the threshold of 67% diversion by the period of 17 years, a yearly increase in diversion of 3.941176% was stipulated. Again, the excess significant digits have no meaning except for reducing round-off errors in the table. For year 9, the required collection from Table 6 was 168.7 kt, and the diversion was calculated as 168.7*0.03941176=6.6 kt. Consequently, the allowed tipping figure was found to be 168.7*(1-0.03941176)=162.1 kt. The dumping rate was not affected and continued as 25.6 kt/year from Table 6. Upon arriving at year 25, the tipping rate reached the threshold value of 33% of collected waste as follows: 218.0*(1-0.03941176*17)=71.9 kt.

sequence of years	popula tion	trash produ ced kt/year	trash collec ted and tipped kt/year	trash not collected and dumped kt/year	trash diverted from landfill kt/year	cumula tive landfill mass kt
0	500000	182.5	146.0	36.5	0	146.0
1	503500	183.8	148.4	35.4	0	294.4
2	507025	185.1	150.8	34.3	0	445.2
3	510574	186.4	153.2	33.2	0	598.4
4	514148	187.7	155.7	32.0	0	754.1
5	517747	189.0	158.2	30.8	0	912.3
6	521371	190.3	160.7	29.6	0	1073.0
7	525021	191.6	163.3	28.3	0	1236.3
8	528696	193.0	166.0	27.0	0	1402.3
9	532397	194.3	162.1	25.6	6.6	1564.4
10	536124	195.7	157.9	24.3	13.5	1722.3
11	539877	197.1	153.6	22.9	20.6	1875.9
12	543656	198.5	149.1	21.5	27.9	2025.0
13	547461	199.8	144.4	20.0	35.4	2169.4
13a	**547461**	**199.8**	**179.8**	**20.0**	**0**	**2274.6**
14	551294	201.2	139.5	18.5	43.2	2308.9
15	555153	202.6	134.5	16.9	51.2	2443.4
16	559039	204.0	129.2	15.3	59.5	2572.6
17	562952	205.5	123.8	13.7	68.0	2696.4
18	566863	206.9	118.1	12.0	76.8	2814.5
19	570861	208.4	112.2	10.4	85.8	2926.7
20	574857	209.8	106.0	8.6	95.2	3032.7
21	578881	211.3	99.7	6.8	104.8	3132.4
22	582933	212.8	93.1	5.0	114.7	3225.5
23	587014	214.3	86.3	3.1	124.8	3311.8
24	591123	215.8	79.2	1.2	135.4	3391.0
25	595261	218.0	71.9	0	146.1	3462.9
25a	**595261**	**218.0**	**218.0**	**0**	**0**	**4673.5**

Table 8. Evolution of landfill mass with threshold targeting from year 9. Bold face values show corresponding situation without threshold targeting.

Additionally, Table 8 provides information on population growth and cumulative landfill mass. The threshold targeting approach allowed for reducing the landfill mass by 26% with respect to the base case of Table 6 in spite of the 19% population and waste production increase during the period. Threshold targeting proved its utility as a management tool.

3.2 Construction and demolition debris (CDD)

In response to a diagnosis, the objective of the study required to construct a management model to efficiently handle construction and demolition debris in the city under study, and to determine the landfill diversion threshold of this waste as an indicator of sustainability.The city did not have a plan for integrated management of CDD. The Brazilian National Environmental Council regulated the handling practices of this material in 2002 through Resolution 307/2002 (CONAMA 2002), but this directive had not yet been implemented in the city. As a consequence, CDD were still collected at assigned locations throughout the city and taken to a CDD dumpsite.

The cited resolution classifies CDD into four categories, namely A, B, C and D. The definitions are as follows.

Category A: Debris that can be reused or recycled as aggregates for construction.
Examples: ceramic components, bricks, concrete, shingles, plasterboard.

Category B: Debris that can be recycled for uses other than in construction.
Examples: plastics, metals, cardboard, glass, wood.

Category C: Debris for which no recycling technologies are available.
Examples: gypsum and related material.

Category D: Hazardous or contaminated material.
Examples: thinners, solvents, oil, paint.

The resolution makes a distinction between small and large volumes of CDD to be taken care of, defined as deposited volumes of more or less than 2 m^3. The document asks for the establishment of networks of receiving points for small and for large volumes throughout the city, for the existence of a free telephone service by which residents may schedule waste pick-up for small volumes with the municipal administration, for the creation of a permanent Sector for CDD management within the administration, for effective supervision of all receiving points, and for environmental education programs directed to the population involved in CDD generation.

According to the document, the receiving points for small volumes have to be fenced, have to provide for separation of incoming waste into classes A, B, C and D, and have to keep records of quantities manipulated. All transfers of material from the receiving points to the CDD landfill are to be the responsibility of the public administration and are to be accompanied by transportation control sheets. Obviously, this standard procedure simply burdens the city's tax payers with all expenses related to CDD.

In view of this questionable procedure, the author's team decided against its recommendation to the city administrators. Instead, they developed a second option more realistic in terms of cost distribution and operating efficiency. The fundamental argument behind this new proposal is that private constructors produce the debris and have to carry the onus of disposal. The function of the municipal administration is to regulate, to supervise and to create the right incentives for private initiative, but not to run the system on tax money. The term "disposal" has to be redefined. The landfill is no longer an adequate place for deposit of CDD. Technology exists for reintegration of class A and B waste into the production chain. The management model needs to address them and stimulate recycling practices within the city. The traditional thinking model, which states that all services are provided free of charge by the public administration, has no place in a sustainable society. The collection and recycling operations have to be run as a business supported by private enterprise. Table 9 relates the cost and benefit distribution at various points in a privately operated system. Figure 2 provides information on material flow in the system.

income items	cost items
Deposit fee at small volume reception stations	Maintenance of small volume areas
Deposit fee at central large volume reception district	Transport from small volume reception stations to large volume district
Sale of class B residues to wholesalers	Transport of class C and D residues to landfill
Sale of recycled class A residues to construction projects	Operation of central large volume reception district and waste treatment
Sale of recycled material to public construction programs	Deposit fee at municipal landfill

Table 9. Economical balance for private waste handling model

The functionality of this model will be explained now. The final destination of collected residues received at any of the receiving stations is a treatment plant for type A residues, the municipal landfill for type C and D residues, and the reverse logistics chain for type B residues. All residues pass through the central reception district where they are separated and their destination is decided upon.

The whole system is operated by a private contractor and has to be financially self-sustaining. Table 9 indicates where the revenues will come from. It also shows where expenditures occur. The municipal administration does not interfere, except that it does buy recycled type A material for public works construction. This item may be negotiated with the system operator as a percentage of total recycled quantity.

All the receiving stations for small volumes are included in the enterprise, such that their operation is the responsibility of the contractor who may subcontract as convenient. All transportation is also part of the enterprise, but may also be subcontracted at will. Referring to Table 9, it is clear that the enterprise has to adjust the receiving fees at the various stations to values that will support the system. It will also have to pay the tipping fee at the municipal landfill. This is an important stimulation for the contractor to maintain a high level of recycle.

All small and large construction and demolition operators have to pay the reception fee at the receiving stations. This is the main new thinking model to be indoctrinated to the community. Heretofore people were used to discard their debris free of charge. The municipal bylaw, which will legally support the model, will have to insist on high fines for clandestine deposits in order to discourage them. The removal of this kind of degree of freedom is the heart of the system. The transportation of small volumes to the receiving stations is the responsibility of the rubble producers.

Large construction companies have to haul their debris to the central reception district where they pay the deposit fee. This central district is the heart of the model. It is at this point that all received material is separated and forwarded to its respective destination.

The number of small volume receiving stations in the city will be decided by the contractor and negotiated with the municipal administration who may rent publicly owned land for this application.

The market is expected to take care of operating details in the system such as the equipment and manpower available at the receiving stations and the central district and the intensity of the sorting procedure. The contract of the system's operator with the municipal administration will set the boundary conditions for the functioning parameters such as capacity adjustments as required and the disclosure of balance sheets to justify the receiving fees.

The diagnosis showed the following contribution of each category to the composition of CDD: A = 75%, B = 15%, C+D = 7% and E = 3%, where E refers to contamination by biodegradable items during the collection process. As materials of categories A and B are potentially recyclable, their sum of 90% represents the diversion threshold achievable by private operators. The management model for the city will be designed to approach this number within an established timeframe, just as was done for domestic waste.

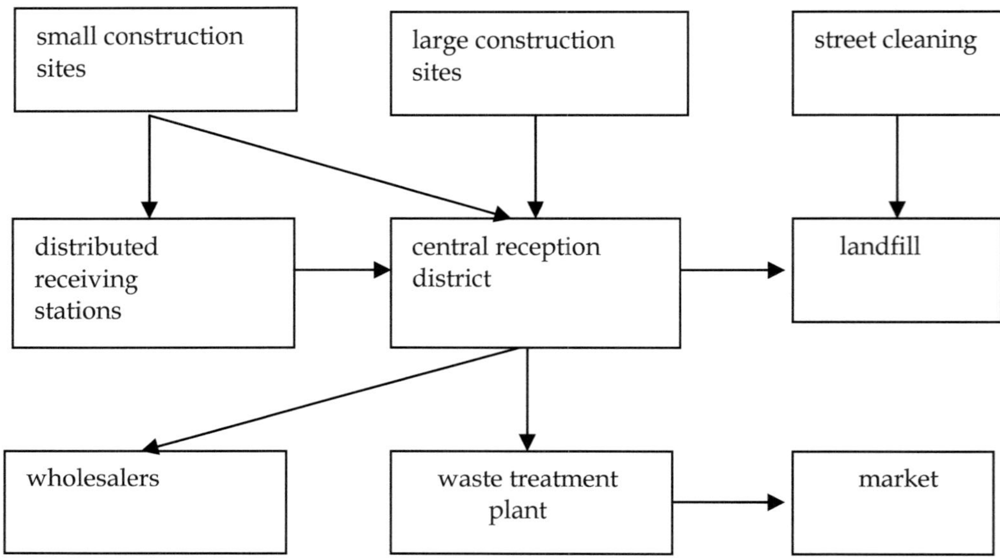

Fig. 2. Material flow of CDD in proposed model

This is a new model, which to the author's knowledge has not yet been experimented with in Brazil. The corresponding bylaw will have to be proactive in the sense that it needs to require constant updating of the management model as experience accumulates. The important fact is that the taxpayer has been relieved from the necessity to support construction and demolition waste handling in town, which is considered part of the sustainability indicator. The starting target of the sustainability indicator measurement will be the complete landfill diversion of type A and B material through privately operated facilities.

4. Conclusions

Urban sustainability has been envisaged as a process, not as a situation.

Landfill diversion of household waste and construction and demolition debris has also been treated as a process that moves towards a target.

The diversion requires a management effort, which in turn also is a process with a final target in mind.

The management talent of the municipal administration has been incorporated into the sustainability indicator.

With a final target in mind and the timeframe to reach it, fractional approaches to sustainability may be reported.

The proposed models induce the city administration to set precise targets on the road to meeting the targets relating to urban sustainability.

The standard management procedure for construction waste suggested by the National Environmental Council has been analyzed and found unfit for sustainable waste management. Its shortcoming is the unrestricted financial burden it places on the municipal taxpayer.

Thresholds have been determined for landfill diversion of household waste as 67% and for construction and demolition debris as 90% of quantities produced. Any target below the threshold can be met by strictly private initiatives if properly stimulated, thus liberating the city administration from financial and physical commitments.

To the author's knowledge, this is the first time such a model is being considered for implementation in a Brazilian municipality.

Urban waste management has been promoted to the position of sustainability indicator for a city.

The threshold targeting method has been shown to be a realistic management tool that can drastically reduce the need for landfill space.

Sample spreadsheets have been provided to show how the management effort may be diluted over time.

Threshold values for landfill diversion are experimental. They are derived from waste composition and source separation tests.

The new concept of sorted waste composition has been introduced and used to identify the diversion threshold.

Sorted waste composition does not reproduce the results expected from raw waste composition.

5. References

CONAMA National Environmental Council, Brazil (2002) *Resolution nº 307 of July 5, 2002.* Establishes directives, criteria and procedures for the management of construction and demolition waste (in Portuguese).

Doyle, R. (1997) Global fertility and population, *Scientific American 276 (3): 20, March.*

Fehr, M. & Calçado, M.R. (2001) Divided collection model for household waste achieves 80% landfill diversion, *Journal of Solid Waste Technology and Management, Chester, 27 (1): 22-26.*

Fehr, M. (2003) See you in 2080 (in Portuguese), *Aguaonline - revista digital da água, do saneamento e do meio ambiente, Rio de Janeiro, Issue 184 p. 1 (2003 11 14)* http://www.aguaonline.com.br (2009 10 .09)

Fehr, M. et al (2004) Proposal of indicators to assess urban sustainability in Brazil, *Environment, Development and Sustainability, 6: 355-366.*

Kanitz, S. (2006) The net national capital (in Portuguese), *Revista VEJA, São Paulo, February 15, p. 20.*

Rees, W. & Wackernagel, M. (1996) Urban ecological footprints, *Environmental Impact Assessment Review, 16: 222-248.*

Report on what makes a good metropolitan city (2004), (in Portuguese), *Revista VEJA, São Paulo, 37 (10): 72-73, March 10.*

Villa, V. & Westfall, M.S. (2001) Urban indicators for managing cities: *Cities data book, The Asian Development Bank pp. 16-36.*
WSSD (2002) World Summit on Sustainable Development, Johannesburg http://www.un.org/jsummit (2009 10 09)
ZWIA (2009) Zero Waste International Alliance www.zwia.org (2009 10 09)

2

Strategic Analysis of Alternatives for Waste Management

Enrique Posada
Indisa S.A.
Colombia

1. Introduction

Cities, large and small, generate sizable amounts of solid waste. Solutions to this problem are many and range from dumping sites and landfills to sophisticate recycling and treatment schemes. This chapter proposes a model for public policy making which considers the cost and benefit of different alternatives to handle the wastes for a time horizon of 20 years. Combinations of several alternatives: Recycling, waste reduction, waste separation, incineration, land filling with lixiviate treating and gas energy recovery, and organic waste composting are considered and modelled and the results presented as a contribution for a more rational approach to policy making than the one currently being employed.

The chapter puts emphasis on the role of ideas and beliefs in the shaping of public policy and in the need to review the belief system of the responsible bodies and of the community, as related to creativity, planning, long term considerations, sustainability, economics, perspectives and view points changes. A review of possible alternatives for solid waste handling is presented, accompanied of some simple models for calculating basic treatment costs, estimate environmental and energy impacts, mass balance flows, resource requirements, labour impact and required time table for rational execution of associated projects.

2. The solid waste problem in cities: a major opportunity for strategic planning and for developing community environmental awareness

The solid waste problem has not been solved efficiently by humankind. Strong forces conspire to make this an awesome problem, among others:

- The consumption habits that dominate modern life, with all kinds of attractive packages for everything and publicity attracting the consumer to increasing expending and to the use of goods that gives rise to amazingly large quantities of disposable materials that have to be dealt with as waste.
- Public awareness that resists the old methods to handle waste and limits the possibilities of landfills and incinerators and that wants waste handling facilities located far away from people's living and working quarters.
- A tendency of the waste materials to be easily mixed and contaminated, following entropy laws of disorder, which diminish severely the recycling and recovery options and makes much more costly to pick, transport, clean, separate and treat the waste.
- A tendency for people to get away from waste handling and recovery, as mixed waste is associated with offensive odors, sticky and nasty textures and ugliness.

- More and more demanding norms, regulations and laws, that limit the options available and increase the costs of waste handling.
- A constant creation of new and more complex materials, that enter the solid waste chain and put severe demands on prevention of damage and avoiding any negative aspects of solid waste disposal, recycling and reuse.

On the other hand, new opportunities arise such as:

- Job creation will be associated with the recycling, waste and recovery processes. Quality work is required to separate, classify, transport, treat, dispose and recover materials and handle waste.
- Creativity will be required to found ways of recovering the materials and treating the waste. Creativity will always be a welcome human trait that will add to happiness and prosperity.
- Technology developments will create jobs and economic success.
- Scientific finds will help to clarify life cycle issues and to diminish fears once solutions are found for the problems.
- Collaboration between sectors and interest groups will help people to live wisely together and to support community projects.

3. Strategic planning

Planning is an attempt to do things in an organized way, trying to define paths to develop the activities related to a given project. A project is the response to a given need and is specified trough objectives and activities to attain them. Planning organizes the activities according to priorities, logical connections between them, available resources and time schedules. It is like a map to travel safely trough unknown territory.

However, unlike perfectly researched maps that describe a well traveled region, projects designed to solve community problems are imperfect and easily fail in their objectives, result more costly or are subjected to harsh criticism. To help alleviate these shortcomings, strategic planning is an interesting tool, as it is designed to consider alternatives, to examine weakness and risks, to take into account strengths and opportunities. In this way, the minds are open to options and to flow with the necessary changes once the project is on the way.

When doing strategic planning, every thing can be considered from two sets of view points. The first set takes a look to a given situation and ask four questions:

- What are the strengths associated to it?
- What are the opportunities associated to it?
- What are the risks associated with it?
- What are the weaknesses associated with it?

These four questions correspond to the classical SWOT analysis (Strengths, Weaknesses, Opportunities, and Threats).

The second set of view points examines the situation in relationship to its state, taking the best possible look to four states:

- The state of the art, which describes how the situation is handled in places or with methods known for quality, evolution and development.
- The ideal or theoretical state, which describes the limits that can be reached when the situation is taken to reversible and near perfect situation.
- The historical state, which describes the situation as it is and as it has been in previous times. This includes the measured and observed state which examines the situation through real observations made with experimental and auditing methods.

- The modeled state, which subjects the situation to simulations and probes, through different simulation techniques, in order to examine how it responds to possible variations.

This four state analysis will be called here STATE analysis. This eight view point examination illuminates the analysis of the situation and permits wiser project formulations. This eight view point exam belongs to a family of view points and analysis techniques that can be used in strategic planning:

- SWOT analysis (Strengths, Weaknesses, Opportunities, and Threats)
- PEST analysis (Political, Economic, Social, and Technological analysis)
- STEER analysis (Socio-cultural, Technological, Economic, Ecological, and Regulatory factors)
- EPISTEL (Environment, Political, Information, Social, Technological, Economic and Legal).

The three last sets of this list correspond in some ways to STATE analysis which proposed here.

4. A simple example of strategic planning

Let us consider a simple situation and let us subject it to strategic planning. What to do in our house with the waste generated by daily family life. We propose this exercise as the initial point of any real strategic analysis of alternatives for waste management in a city. As far as we know, it has not has been done so far with the participation of a representative number of households and, not surprisingly, there is a lack of awareness in most people about how to deal with this situation in a responsible way.

The following matrix performs a SWOT analysis (Strengths, Weaknesses, Opportunities, and Threats) to this situation

Strengths	Opportunities
Available spaces to handle waste, such as yards, basements, cellars and gardens. Education and knowledge Creativity Consciousness Family structure and civic beliefs and sense of compromise Small size Relationships with neighbors and friends Free time Surplus money and savings available for small and reasonable projects Funds and support available for projects at the family and community level Existence of community groups and associations Existence of support from waste handling and municipal authorities	A family solid waste project could unite the family Savings by recycling and rationalization of consumption habits Cleaner spaces and better habits Better communication through eating and acting together and interacting Better education of children through real life examples Acquiring a sense of pride and self esteem New methods could be discovered and shared with friends and neighbors Community projects could be developed to better barrios and neighborhoods Organic fertilizers and materials could be developed to improve gardens and save money The city waste load and waste handling cost could diminish

Weaknesses	Threats
Lack of communications Lack of vision and family purposes Distractions and lack of time to share and work together as a family Poverty and lack of means Cynic beliefs and pessimism about civic values Fixed ideas Lack of awareness about the environment and about the importance of resources and of good housekeeping Lack of time Lack of leadership Lack of knowledge about what to do Lack of compromise	Projects could fail and cause discouragement Family projects and community ideas could fail because of lack of municipal support Public health risks when waste is mishandled at the household or neighborhood level Projects could be discouraged by waste collection practices and regulations Disputes and opposition of neighbors to recycling and to communal practices for waste handling Projects could be abandoned for lack of vision, for practical difficulties or for lack of means Presence of molds, bugs and animals attracted by poorly handled waste

The following is a proposed STATE analysis, described in a colloquial and somewhat poetic and literary way.

The state of the art in household waste handling

I researched and researched looking for the perfect house,
in which mother and father and children worked together
to protect the earth from the themselves that throw things away.
to learn from each other and also teach things away.

I then saw this mother of three who prepared food lovingly
and invited all together to the common table to share and to eat with humble passion,
so that not leftovers, ever, were left,
and only prayers and good humor were thrown away every day.

I then visited their simple garden, a little space of flowers and essences,
where the children learned that nature can be created by men
where they learned to perform the miracles of composting
and witnessed the good ways of the earth worm.

Of course there were things that were discarded and that had a bad smell,
and useless plastics and bags, plus disposable vases and dishes,
dirty napkins, toilet paper, all the trappings that make easy modern life,
but they somewhat managed to handle that in ways surprising and wise.

To begin with, there were few of these things, and well classified and separated,
in clean tidy places well marked and adorned with colors and drawings,
so that when they went back to the cycle of life, dead or alive,
nobody received them with contempt, but with affability and gratitude.

Unfortunately I must tell that this special place only exists in dreams.
However, I am working on it, little by little and it is coming,
as you, reader will feel the challenge and take care,
as you, men and women of households will willingly dare.

The ideal state

The following list (table 1) describes ideal situations for several kinds of household generated waste. The reader should compare it to his or her own case and add new possibilities. Cities should prepare ideal models of behavior with the help of committed families and persons, so that citizens see clearly se how they can contribute to the diminishing of the waste load to be handled by the cities.

The historical state

Yesterday I was walking in my neighborhood observing waste handling, and this was what I saw. First of all, my city is known as the city of flowers, but there were vey few flowers on the windows of the houses and I saw no pots with flowers in the sidewalks of the streets. Almost everything was quite gray and covered with dull cement. Most sidewalks were clean, because the humble ladies and men employed by the cleaning authorities just passed by and cleaned the cumulated debris, as shown by the black plastic bags packed with waste and piled by the streets, waiting for the municipal truck to pick them to be taken to the dumping site, forty kilometers away. I saw some people putting domestic waste bags on the sidewalks, mostly used plastic supermarket bags full of mixed refuse, some loose, some closed. Some of the materials spilled from the bags, and I though that streets dogs could cause a mess, but, luckily, I did no see any.

Then I saw this lady, carrying some bags with her, the lady of the recycling. She and some men of her kind, walked by the streets, opening some of the bags, here and there, taking some plastic bottles, cardboards, paper and, from time to time, a piece of metal. They were careful, but some almost unavoidable spilling occurred. I felt uneasy, because I like clean streets and clean sidewalks, but did nothing about that. Then the waste truck came. A big truck, imported from USA, somewhat noisy and imposing. It made its way through the cars parked and running, unbelievably finding enough space to pass. Two men walked with it, picking the bags and throwing them inside. Some of the bags exploded when the mechanism pushed them in and I saw one of the men receiving drops of smelly liquids in his face coming from one exploding bag. Some twenty meters away, there was a little park and several open waste bags piled in front, with two gallinazo vultures pecking the bags, looking for some hidden manjar. The workers had to frighten them to take the bags to the truck.

In one of the houses, just in front, a young woman was sweeping from the inside of the house to the sidewalk, but it was not too terrible, only some dust. I also saw a car driver pass by and throw a cigarette butt on the street. This attracted my attention to the many butts on the floor. Then, the rain came. It was one of those tropical sudden rains that cause rivers of water on the streets. Luckily, I was near a covered sidewalk, in a bus stop, and waited there until the rain stopped. The water in the streets began to flow, dragging all kinds of things: leaves, bags, plastic cups. The things, really not too many, were however enough to cover the grate of the nearby culvert which made the entering water jump and I had to move so that my feet did not get wet. Then the rain stopped and I went back to normal life, away from the waste truck, to go on, although a little more aware of the people that have to pick the waste we do not know how to handle yet. I promised my self to do something. I started by writing this little story.

Kinds of waste	Proposed ideal handling situation
General	Minimizing waste and good handling practices: • Separate waste and keep it in several marked clean, covered and separated recipients to maximize opportunities for use and treating. • Support good practices with the neighbors and participate in communal waste handling projects. • Minimize waste through good buying habits. • Stimulate simplicity, prudence and avoid habits of excessive materialism and waste. • Keep some simple registers of amounts of waste, from time to time, seeking waste generating minimizing goals. • Be aware of city programs to minimize and recycle waste and support them with words and behavior.
Food leftovers	Not leftovers resulting due to good cooking practices and good eating table practices: • Loving cooking which takes into account people habits and health. • Family table ways that encourage communal servings from food sources and only taking what can be eaten, without squandering the food • Not eating in TV rooms or bedrooms. • Good buying and storing practices to avoid food to spoil. • Good eating habits to avoid excess weight or eating disorders
Organic leftovers resulting from vegetables and plants	Minimizing leftovers through good buying and cooking practices. Collecting them in several marked clean, covered and separated recipients to maximize opportunities for use and treat. Aerobic composting for use in house and community gardens. Look for the city to develop systems to process separated organic waste to generate electricity or agricultural products.
Garden waste and cuttings	Establish some kind of aerobic composting system to use all the materials generated. Do not burn or throw away materials
Plastics, glass, paper and metals	Make sure all materials of this kind are recyclable Know the kinds and separate and keep them correctly and make sure they are handled to somebody dedicated to recycling or properly treating them Get acquainted with persons and groups dedicated to handling these materials and support them Use the municipal facilities available for recycling and contribute to their cleanness and good use
Used clothing, apparel, furniture and books	Share used items with friends, relatives and neighbors. Get acquainted with persons and groups dedicated to handling these materials and support them. Do not accumulate items without need

Table 1. Ideal household waste handling situations

The modeled state

As it fits to an engineer, its is nice to calculate things and design small and simple models to determine what will be the effect of several actions, in this case, ones that a citizen could perform, for the benefit of the waste handling problem of a city. Two lines of actions were simulated, as follows:

- Keep doing what it is currently being done.
- Starting to work and influence people: Give lectures on good practices to minimize and handle waste; change the behavior in the house and in work to a perfect behavior as proposed in the idea state analysis; Get involved in projects related to waste problems.

The model is designed in a simple way, as described by the following graphs and tables. It was applied for a range of 20 years for a population of 1.000.000 people.

Variable	Units	Range
Cost of land fill disposition and residue recollection	US $/ton handled	15 - 42
Value saved by recycling material	US $/ton recycled	25 - 70
Quantity of waste produced daily (initial)	Kg/person	0,80
Persons that are influenced by direct action (Primary influence factor on people)	Persons per year	0 - 500
Persons that are influenced by indirect action through primary influences (Secondary influence factor on people)	Persons per year	0 - 50
Factor of yearly change in patterns of residue generation (1.0 means no change) followed by influenced people		0,1 - 1,0
Factor of yearly change in patterns of recycling (1.0 means no change) followed by influenced people		1,0 - 2,0

Table 2. Variables used for simple model of the effect of personal actions

Figures 1, 2 and 3 show the general behaviour of costs under several suppositions related to recycling rates, disposition costs and recovery costs due to recycling.

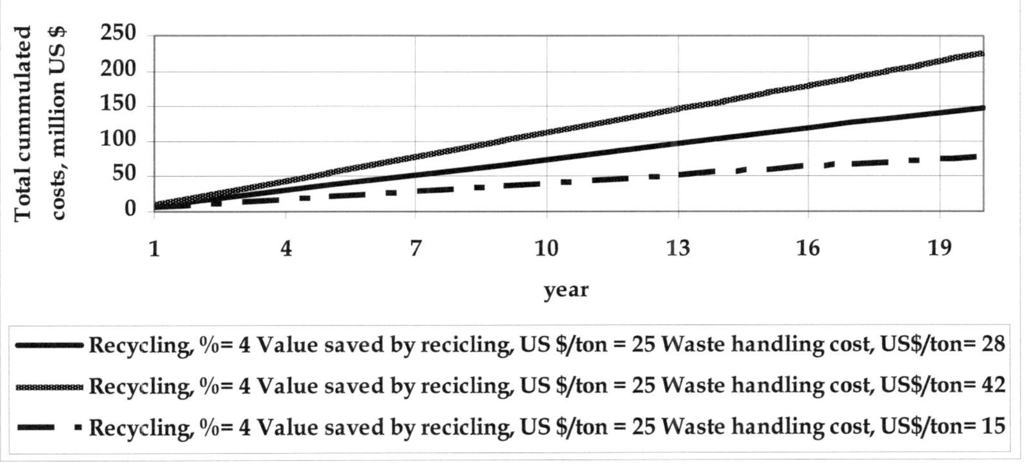

Fig. 1. Total accumulated costs at various waste handling costs

They allow having a general sense of the situation, which is quite useful when modelling something complex. Figure 1 shows that waste handling costs have a very large direct influence in the analysis.

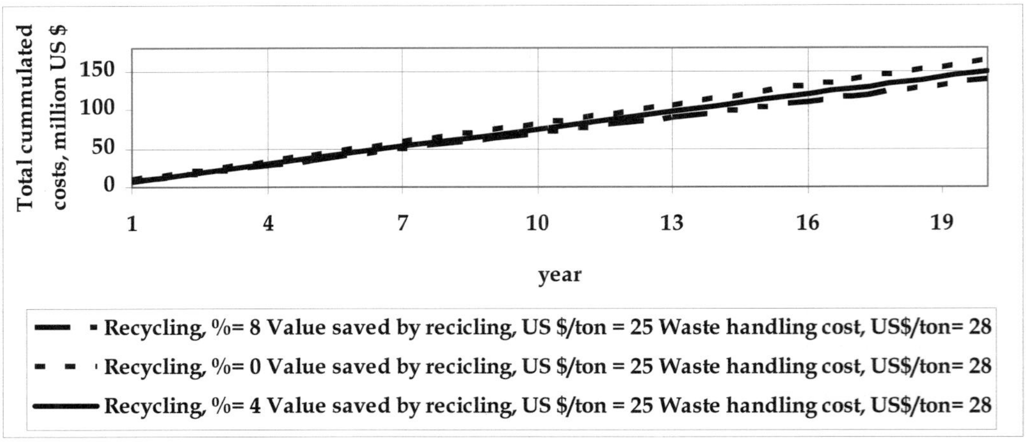

Fig. 2. Total accumulated costs at various recycling rates

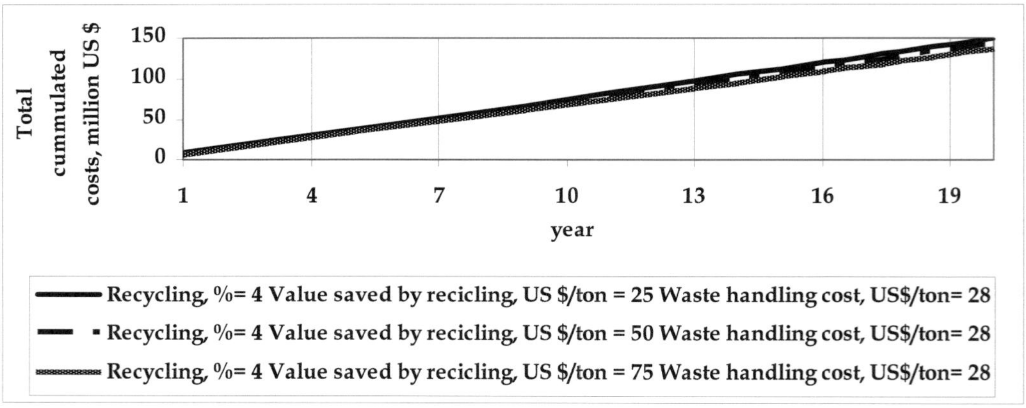

Fig. 3. Total accumulated costs at various values saved by recycling

Non-dimensional analysis is an excellent tool to understand situations in a more general way, as shown by figure 4. It shows that increased recycling and increased added value of recycled materials will mean a powerful combination to save money in the waste handling problem.
Now the model will simulate the effect of the actions taken to influence people, so that the influenced persons will do two things: change their patterns of residue generation for the good (less generation) and help change the recycling patterns with their influences. But also by having a secondary influence in other people (no tertiary influences were considered).
Figures 5 to 9 show the results of these simulations. They show that influences can be very powerful, especially when they are directed to change residue generation patterns. The percentage influenced can be very large if the actions are able to reach 500 people every year and if each of them influences also 50 other people. In this way in 20 years the influences will reach 51 % of the total population. This scenery, of course, is quite optimistic, but no

impossible, as shown by the effect of people like Rachel Carson. A more probable influence will be that of 200 direct influences per year with 20 secondary ones.

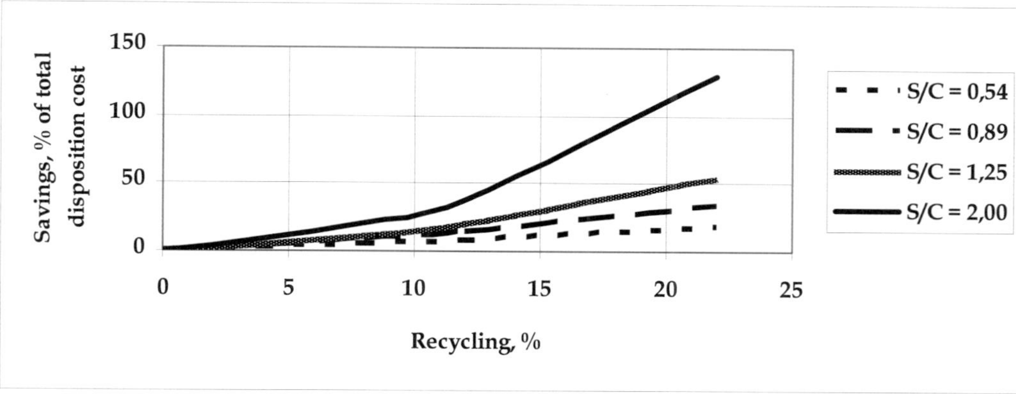

Fig. 4. Savings (Percentage) of disposal costs) as function of recycling rates for various ratios of value saved by recycling (S) to waste disposal costs (C)

Primary influence factor on people	Persons per year	0	500	200	200
Secondary influence factor on people	Personas per year	0	50	20	20
Percentage of population participating under primary and secondary influence after 20 years		0,0	51,0	8,4	8,4
Factor of yearly change in patterns of residue generation		1,00	0,95	0,10	1,00
Factor of yearly change in patterns of recycling		1,00	1,25	1,00	2,00
Cost of land fill disposition and recollection	Million US $	155	151	148	154
Net cost after 20 years	Million US $	149	145	142	148
Value saved by recycling after 20 years	Million US $	5,76	6,26	5,50	6,07
Recycling initial	%	4,00	4,00	4,00	4,00
Savings initial	%	3,86	3,86	3,86	3,86
Recycling after 20 years	%	4,00	4,44	4,00	4,22
Savings after 20 years	%	3,86	4,33	3,86	4,09
Savings versus no action	Million US $	0,00	4,314	6,708	0,665
Resulting change in recycling patterns after 20 years	%	0,00	10,97	0,00	5,44
Quantity of waste produced daily per person after 20 years	Kg	0,80	0,77	0,74	0,80
Resulting change in generating patterns after 20 years	%	0,00	4,01	7,61	0,00

Table 3. Example of results of simple model of the effect of personal actions

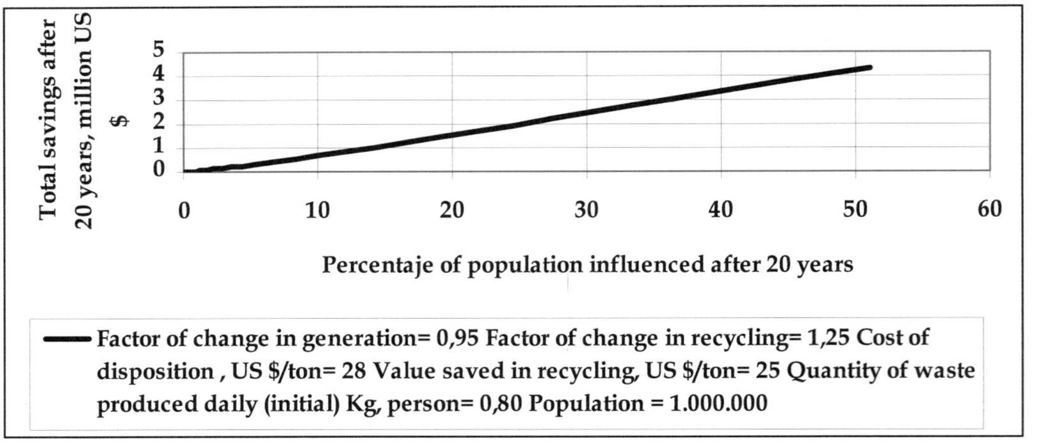

Fig. 5. Savings as a result of influences in population behaviour

Savings could reach significant values of the order of several US $ million dollars for the case simulated. This calls for the importance of cities to consider stimulating individual actions.

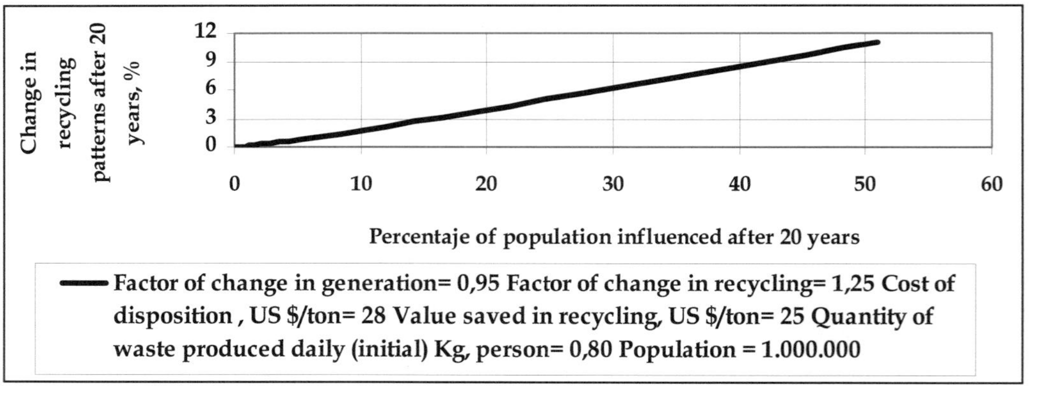

Fig. 6. Change in recycling patterns as a result of influences

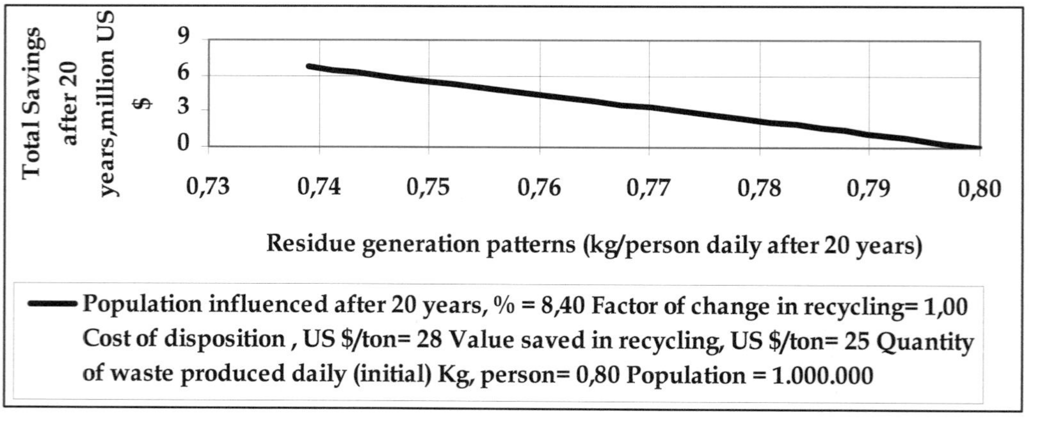

Fig. 7. Effect of the generation rate on total savings

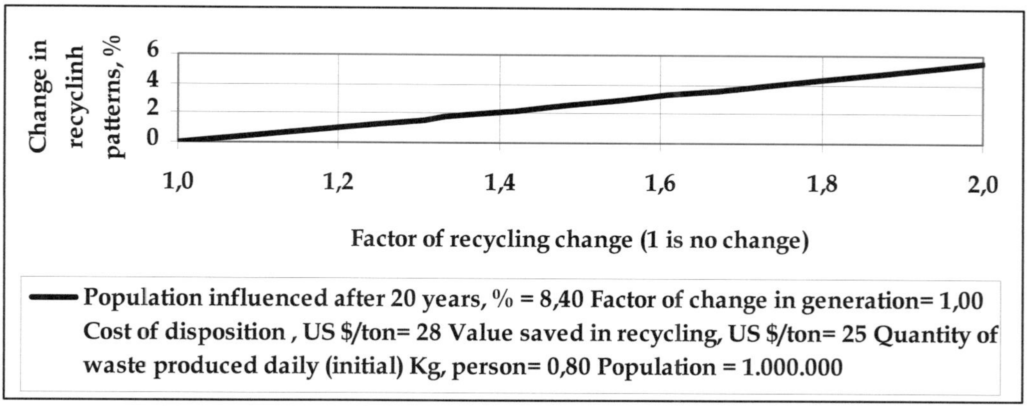

Fig. 8. Effect of the factor of recycling change on the variations of recycling rates

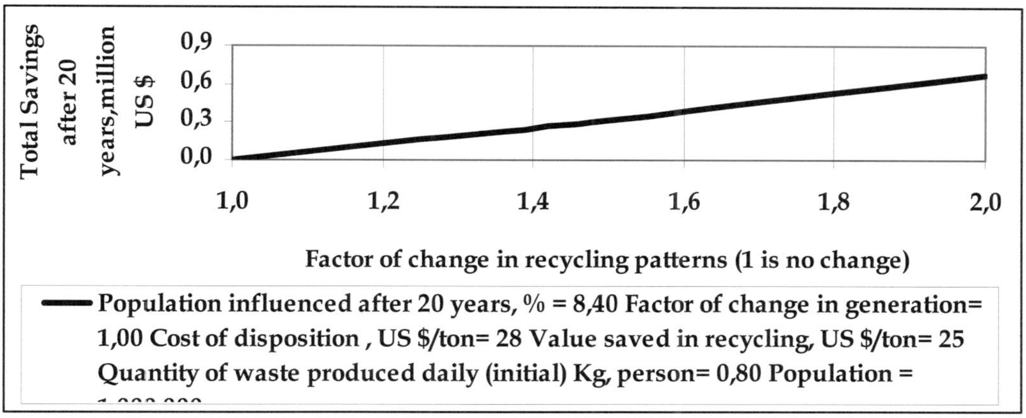

Fig. 9. Effect of the factor of recycling change on the total savings after 20 years

It is concluded that strategic analysis permits to observe the effect of alternatives and actions in a simple and powerful way.

5. The role of ideas and beliefs in the shaping of public policy

Creativity should be a major tool to resolve environmental issues, especially when dealing with recycling, recovery and re-integration. Major obstacles to go in the right direction are related to fixed ideas in the minds of people. Ten principles of creativity are now presented which have the potential to awaken people and to open their minds towards right actions and positive attitudes. Practical ways of application are suggested

The need for a creative approach and where to find it

These basic principles of applied creativity can be discovered and appreciated based upon one's intuition, personal experience, and studies. They possess great potentials that can be explored and, specifically in this case, can be applied to the development of creativity in the recovery and recycling of by-products and waste. When pondering these subject matters, it is helpful to first see oneself as an ethically responsible creator of new realities and, in this particular case, as an author of the reality of sustainable development; then, these principles can be successfully applied.

The "no-locality" principle

In the heart of all things there exists a spark, a sense of divinity. We become creative and effective when we discover our duties as creators inspired by our divine sparks. This principle might be applied by first declaring that things must not be labeled as good or bad, since everything has a value and a divine spark, which is non-local and non-limited. A contaminant, a by-product, or a waste is a substance waiting to be correctly labeled as a source of opportunities for changes that could be translated into benefits. This leads to the idea of considering the controlling of environmental pollution, and the recovery and recycling as profitable and attractive processes. Such vision will facilitate the optimization of industrial processes, making them more profitable and less expensive as fewer by-products are generated. This will also allow for savings in materials and energy, less but better maintenance, and increased personnel commitment with better sense of belonging to the organization. Therefore, the challenge is to find the hidden benefit in any environmental handling by discovering the "good" zone that exists in any "bad" process and turning a threat into an opportunity.

The "participation of the observer" principle

The enormous wealth of our ecosystems provides an opportunity for discovering new ways of living together with active community participation. We are part of a lively laboratory of complexity, in whose foundation the secrets of life vibrate. Everything we discover is deeply influenced by the motivation that leads us into investigating a given situation. In other words, the observer creates the "reality". Correspondingly, we can reach a sustainable future if we focus on a creative vision.

With our participation, we can create local industries and establish teams dedicated to waste cleaning and to the recovery and recycle of by-products. In this way, the quality of life for the poor people dedicated to the humble tasks of recycling waste can be notably improved. New uses for garbage, concurrent with new ways of diminishing and cleaning it, can be found through teamwork and common interest.

The "uncertainty" principle

It is not possible for facts or aspects of a given situation to be completely true. Life is an indeterminate work in progress. In the absence of certainty, a favorable change is always possible. Handling the relations with the workers with love and openness, so that they feel stimulated to contribute with useful ideas to the recovery, minimization, and recycling of waste. Learning to hear the complaints from the community in order to see them as opportunities for improvement, for growth and for development of new technologies, instead of seeing them as insurmountable misfortunes or nasty products of the production system. Visualizing laws and control rules as signs of alertness to changes and adaptation to better and more efficient realities, as means for the communities to handle uncertainties.

The "complementary vision" principle

Knowing the existence of duality in all things liberates us from creating artificial limitations. This freedom makes responsibility possible. This is the principle of "duality", or the "complementary vision". It is the vision that accepts that there are many ways to solve a problem and that it is not necessary to judge or fix arbitrary limitations on things or persons.

Superior levels of consciousness are characterized by an ample vision of the reality, where different viewpoints abound. In this way, the observer should witness without criticizing and be convinced that everything complements itself in one way or another. It is always possible to improve; it is always possible to evolve and to optimize. What seems to be perfect and simple may have some not-so-good aspects. Discovering them is an important step to improvement. Cost-benefit and strategic analyses are valuable tools, as they allow to see the complementary aspects of the problems and their solutions. Forming working teams in the companies, at all levels, and also among different companies, is an excellent way to achieve harmonic and complementary work.

The "unity" principle

Everything in the Cosmos is inseparable. From the union of concepts, of bodies, of minds, of ideals, of efforts, of information, of all things, creativity arises. This principle suggests the idea of unity and of the possibility for discovering it. Unity can be attained through "feeling" things, people, ideas, and objects in general as close to oneself as possible. To do so, one takes any object (person, belief, concept, rule, problem) and puts oneself in the place of the object to feel what it might feel being so. By doing this, we can become intimate, compassionate, and close to any object in the entire creation. This is a miraculous and effective way to discover unity. Once one can "feel" a waste or a by-product, one will know what to do with it.

The "parallel universe" principle

There are several levels of reality. There are several levels of consciousness. This is a principle that opens the door to hidden realities, to the unknown side of things. It refers to revealing ourselves as beings with various levels of consciousness, which we could awaken in order to experience them and to enrich our activity.

When we envision and make declarations for a new reality, which are stated clearly in first person and in the present time, creative energy comes unbounded and applicable, resulting in profitable and interesting ideas that arise ready to be developed. This creative state can be attained after following the steps described below (which are basically tools of strategic analysis):

- We identify a problem to be solved and define it clearly.
- We declare explicitly a target that identifies the envisioned solution. This is done at both personal and group level.
- We elaborate a personal, or group, declaration that is clearly stated in present tense and in first person.
- We move on to propose and take actions, with the certainty that there are going to appear synergies, help from others, new ideas and proposals that will line up in the direction of the sought target.
- We stay alert and ready to apply and stimulate those synergies, ideas and proposals.

The "relativity of time" principle

The limited and linear concept that we have assigned to time prevents us from seeing the Universe at a glimpse. Because of this, we only learn little by little. Life is more beautiful when we understand the eternal present at all times. Here we can find tools to handle one of our biggest limitations: our beliefs that we have not time, that we cannot change things, and that dead-lines are fixed. With these constraints, we tend to feel harassed and easily give up. We tend to leave things to be done at the last minute, and avoid preparing and following timelines, thus lacking a global viewpoint. It is recommended to elaborate and continuously update lists with pending matters that need to be resolved. It is appropriate to estimate the necessary efforts required to find possible solutions to problems, in addition to their cost and benefits, so that later only those projects that turn out to be attractive are proposed.

It is important to have the mentality of insisting on the proposal of good ideas until they become practical. Projects should be managed on a budget, using timelines and with consideration of all possible feedbacks. Once a project is executed, consistent operation should be maintained until the awaited results materialize. The investment in human resources dedicated to the handling of waste minimization will lead to real and profitable projects. The establishment of a Research and Development department, with environmental responsibilities including process optimization, is profitable even in relatively small companies.

The "energetic field" principle

Life, nature, and all of us are a manifestation of an infinite energetic potential. We are part of an incessant flow of mass and energy. Obviously, there is space to explore the wide, almost unlimited, zones of energetic field in order to find new niches and creative contributions. But energy fields transcend and spread from what we know in the material realm to the totality of human realities, connecting all ways of consciousness in a mysterious way. For example, we have begun to know "technically" what we have always known from our hearts: that the forces of love exist, that they have physical effects, that solidarity, kind attention, and appreciation are the magic ways of composing reality. The establishment of networks of solidarity, the stimulation of cooperative systems, the support of public-spirit and of all the manifestations that enrich community work and brotherhood will be the sources of employment and happiness. Poverty, waste, and violence are the results of selfishness and lack of solidarity.

The "entropy" principle

Everything is in a process of either breakup or integration. Crisis contains the keys to development. Things agitate to change their level. From the apparent disorder, the miracle of dynamism and self-reference is born. We must not be afraid of change.

The creative handling of change and crisis has always been an opportunity to grow. This calls for a very special belief system, full of confidence in the capacity of people and society to respond. Leaders show their class in moments of change and crisis. The formation of leaders is an excellent investment for a society that wants to wake up. Leaders are beings that believe in their divine spark and in those of others, and they dare to get out of their average comfort zone. Without leaders, the society is left with chaos, unresolved conflicts, and negativity. This is "entropy in action". Waste and pollution are signs of entropy and lack of leadership.

The "infinite underlying variety" principle

There are powerful effects hidden in small changes that can influence reality. The chaotic order is a natural part of the existence. The reality is much more complex but, paradoxically, simpler than we believe. The celebration of complexity is the ceremony for initiation of teamwork. We are not forced to experience oppressive or fixed realities, since there are powerful effects hidden in small changes that can influence reality. Existence can always be catalyzed and unexpected things can surge. This applies to wastes since they constitute realities that can be changed.

The role of ideas in the developing of environmental awareness

There was a great North American poet who gave rise to a literature current that opens consciousness towards nature: Henry David Thoreau. It began for him as a personal experiment. During two years, from 1845 to 1847, he lived deliberately, by his own means, in a natural zone, and that inspired him to write its classical poetic work, Walden, in 1854. Nature writing is one of the major innovations of American literature, which also includes Rachel Carson, the initiator of modern environmental awareness. Other significant figure is Ansel Adams, the American photographer and environmentalist, well known for his beautiful photographs of the American West and Yosemite National Park.
It is amazing how much the environmental movement owns to them and, at the same time, how little their subtle methods for consciousness development though images, poetry and inspired writing are employed in actual environmental activities. Illuminated by these great pioneers, whose knowledge of the living world, based on experience, was refined and projected as images, poetry and literature, people should take a look to the situation from a different perspective: to use poetry and images as a cradle to generate idea and attitudes, which in turn, will generate experiences.

Two major levels of consciousness: A list of possibilities to experience and to understand nature and the world

Human beings need an open mind in order to transform their present form of interpreting and experiencing nature, if we are to live in a sustainable civilization. This applies to everybody, as it is everybody who must participate in the necessary change, not only the illustrated elite that aligns with the environmental movement. These new viewpoints require a more complete and integral use of the entire nervous system, which will be able to integrate two necessary aspects: scientific observation and lyrical or poetic expressions.
The following table shows a list of two major possibilities to experience and to understand nature and the world. Both are important and necessary.

Aspect	Logical	Poetic
Dominion	Mind - The conscious	Body - The unconscious
Perception	Thoughts – Sensations	Feelings – Intuition
Memory	Words - Numbers - Parts - Names	Images - Faces - Patterns - Global
Expression	Verbal – Oral – Counting - Writing	Nonverbal – Gesture - Drawing-Pothook
Thought	Analytical - Linear - Logical - Rational Sequential - Convergent	Visionary - Spatial - Analogical - Creative Simultaneous – Divergent
Action	Probing - Executing	Visualizing - Projecting
Organization	Norms – Capital - Resources	Vision - Values - Motivation
Definition	White and black – Sure - Assertive – Clear - Direct	Gray or colors - With alternatives - Suggestive and integrative – Indirect

Table 4. Two major possibilities to experience and to understand nature and the world

6. A review of possible alternatives for solid waste handling

The importance of good administration and sufficient human resources

It could be that the entities responsible for waste managing in a given city, may have a lack of adequate technical administration and resources. It is important to count with sufficient human resources to focus on the appropriate technical handling of the solid waste. There is a tendency for these organizations to work mostly with independent contractors, lacking enough coherence and integration. Frequently the internal workload is too large for the few people who have technical responsibility and this gives rise to attention problems. It is important to have clear technical procedures, technical know how and human resources, continuity in the activities. It is important to have a clear capacity to plan, to study, to make tests and monitoring, to pursuit and develop ideas, to optimize the processes and to introduce changes, to respond to environmental authorities, to the communities and to mass media and to handle the contractors.

The importance of internal planning

As the solid waste problem is so complex and things are changing due to population variations, material variation, regulation changes and market forces, it is very important that the entities responsible for waste managing in a given city, have some kind of integral planning program. It could be that the organizations tend to act in reaction to daily pressures in the middle of many limitations. It is important to count with an integral approach based on goals, backed by enough dedicated personnel, monitored by management indicators, rich in good communications with the authorities and the community and instrumented with cost benefit analysis to rationalize costs.

Integral planning model as a basis for strategic analysis

A program of this type should be able to consider a complete set of alternatives, including the following situations:

- Program of education, separation in the source and recycling.
- Adequate collecting and transportation of waste
- Plant for recycling plastics, paper and metal.
- Electronic waste handling
- Organic waste handling, use and recovery
- Stations for waste handling
- Design, planning and administration of appropriate land fill sites and waste disposition facilities
- Handling of lixiviates
- Treating of biogases generated
- Thermal plant for recovering energy,
- Incinerating
- Design, planning and administration of appropriate land fill sites and waste disposition facilities

7. Modelling clues and helps

Composition of waste and estimating value in a general way

The composition of the waste is a very important strategic variable. And it is a variable in itself, because it changes in time and it will be distributed in different ways according to the district considered. The table shows the "average" distribution for a large city in a developing country.

Material	Percentage
Organic	53,22
Paper	14,62
Plastic	10,51
Glass	4,87
Cardboard	3,14
Textile	3,00
Construction rubbish	2,89
Metals	1,86
Bones	1,37
Leather	0,80
Other	3,72
Total	100,00

Table 5. Considered weight distribution of municipal waste

Average values are used for general planning, but variations should be considered. Future general distributions will resemble the current distributions for more prosperous districts. Some kind of model should be drawn to understand the distribution within the city according to the economic status of the population in the different districts.
The composition should be looked at from the point of view of recycling and use of the materials. It is advised to give a value to the materials, to have in mind that they really are valuable items. The following table reflects this approach, for the weight distribution of table 5. This will help to vision the possibilities hidden in recycling schemes, as compared to just putting the materials in a land fill in a non discriminated and arbitrary way.

Material	Estimated range of value, US $/kg	Average contribution to total value, $/kg total waste	Contribution to value, %
Organic	0,01 - 0,10	0,029	15,58
Paper	0,025 - 0,25	0,020	10,70
Plastic	0,25 - 1,75	0,105	55,93
Glass	0,025 - 0,30	0,0079	4,21
Cardboard	0,025 - 1,50	0,0027	1,46
Textile	0,010 - 0,025	0,00053	0,28
Construction rubbish	0,005 - 0,015	0,00029	0,15
Metals	0,50 - 1,75	0,021	11,14
Bones	0,025 - 0,050	0,00051	0,27
Leather	0,010 - 0,025	0,00014	0,07
Other	0,005 - 0,015	0,00037	0,20
Total		0,188	100,0

Table 6. Proposed method to value residues in a general way

Collecting of waste

This is a critical part of the strategic analysis. Garbage trucks are very expensive and collection routes complicated. Linear programming or a similar kind of technique should be used to better understand and plan the routes and the transportation needs. The following table shows a typical cost distribution for a waste truck, which usually account for more than the 50 % of the waste handling costs.

Fuel consumption	km/gal	3,70
Cost of fuel	US $/gal	2,8
Travel distance	km	152
Waste load	ton trip	12
Cost of fuel	US $/km	0,76
Cost of fuel	US $/ton	9,58
Filters and lubricants	% fuel	20
Filters and lubricants	US $/km	0,15
Filters and lubricants	US $/ton	1,92
Spare parts and maintenance	US $/km	0,80
Spare parts and maintenance	US $/ton	10,13
Tires	US $/km	0,25
Tires	US $/ton	3,17
Financing and depreciation	% monthly	1,56
Investment	US $	160.000
Trips per truck monthly		25
Financing and depreciation	US $/km	0,65
Financing and depreciation	US $/ton	8,30
Labor	hr/km	0,158
Labor	US $/hr	4,5
Labor	US $/km	0,71
Labor	US $/ton	9,00
Insurance and taxes	% monthly	0,43
Insurance and taxes	US $/km	0,65
Insurance and taxes	US $/ton	2,29
Total	US $/ton	44,38

Table 7. Model for waste collection and transportation costs

Economics of recycling

Recycling has been a growing industry and could contribute greatly to create employment, a more sustainable use of resources and prosperity, if it is seen at from an integral point of view. In many cases, it is left to the spontaneous action of poor and unemployed citizens or to the operation of informal enterprises. Many of those people undertake recycling under hard conditions, without benefits or decent working environments.

Plastics, which as seen in table 6 are the more valuable resource in the waste basket of a city, can be recycled in many ways. Recycling units for plastic are commercial in the international market and could also be developed with local creativity, design, engineering and construction. It is important to create value with them, by manufacturing products. In this way, prosperity and jobs will be created locally in a very responsible way. Of course it is fundamental to stimulate separation in the source, as compared to separation after the plastics are mixed and dirtied by the mixing, especially with organics. In order to visualize

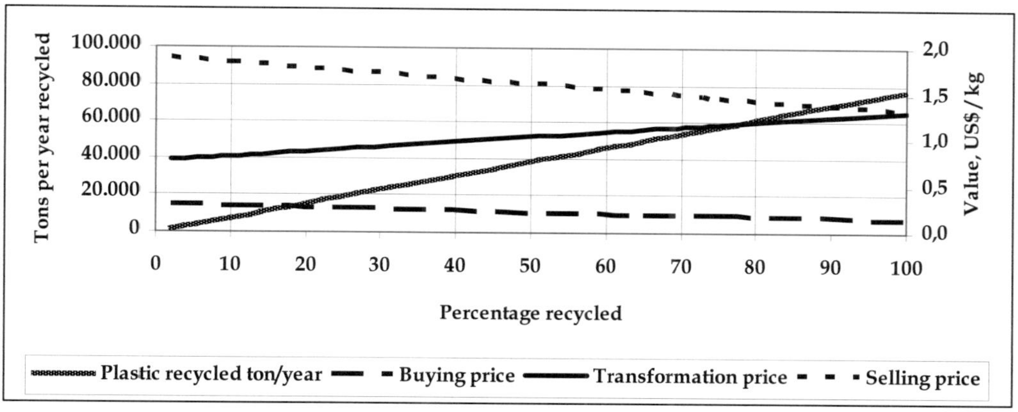

Fig. 10. Example of economics of a recycling scheme. Prices as function of recycling capacity

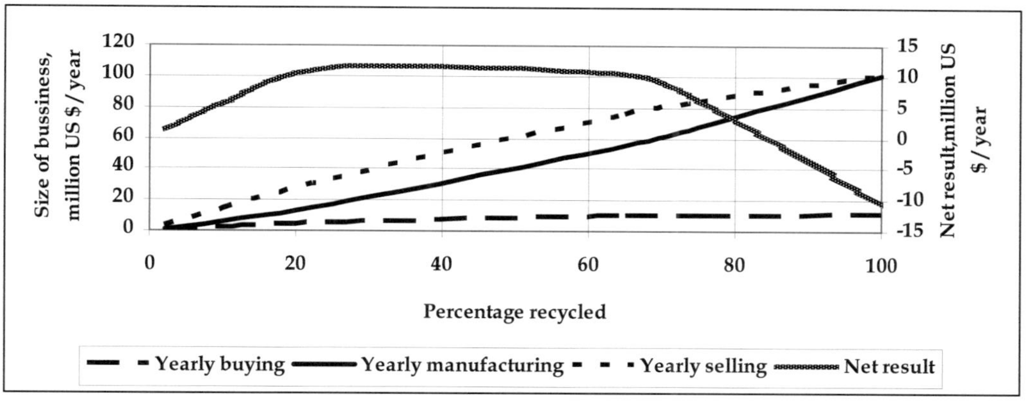

Fig. 11. Example of economics of a recycling scheme. Size of the business and recycling

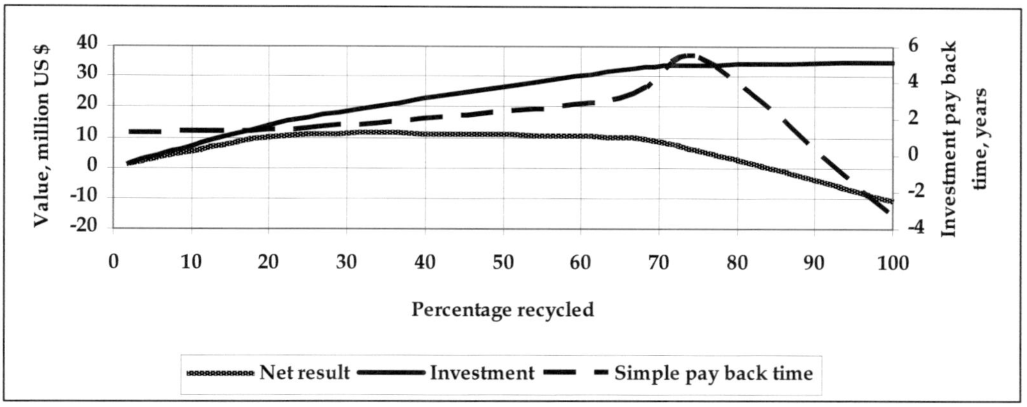

Fig. 12. Example of economics of a recycling scheme. Net results and pay back time.

opportunities, the following analysis, based in a real case of a production plant for recycling plastics and converting them to final products (plastic wood, pellets, bags) with a capacity of 14.256 ton/year and a new investment of 12,5 US million $ is shown in the next graphs. The analysis was made for a city producing 2.000 tons per day of waste, with 10,51 % in weight of plastics.

Schemes like the one shown can be applied to each resource available in the waste basket of the city. This is responsible planning, which will lead to actions.

It is very important to include electronic waste handling in the strategic analysis. The cities should negotiate some type of returning scheme with the companies that manufacture electronic items. A preferred alternative is to include local separation and added value facilities as part of the scheme, supported by the manufacturers.

Organic waste handling, use and recovery

It is mandatory to have a high level of consciousness and give special considerations to organic waste, especially for cities that generate high percentages of organic waste. Organic waste can be easily converted in valuable materials for agriculture use. However it is not the case in general. This is due to fixed beliefs systems that put the emphasis on the false easiness of mixing everything and put it away. The old countryside ways were lost entirely when people migrated from the rural areas to the cities and dumping everything to the garbage truck became a necessary modern feature. It is important to understand the values contained in the organics and to have several sceneries to simulate, for example:

- Schemes to stimulate good practices at households and separation of organics.
- Schemes for household and residential units recycling to make compost for use in gardens. Also larger schemes to process separated organics at landfill sites, especially material from institutional sources.
- Schemes to have special recollection of organics to be taken to processing facilities, both wet and dry, to generate biogas, electricity, agriculture materials and even new products, such as biogas, alcohol, hydrogen and biodiesel.

Handling of lixiviates

Lixiviates are going to be generated in any landfill facility as a result of the process undergone by the organics and the effect of rain. The following figure shows a real situation of a land fill site located in a place with very high precipitations (3.600 mm per year). The

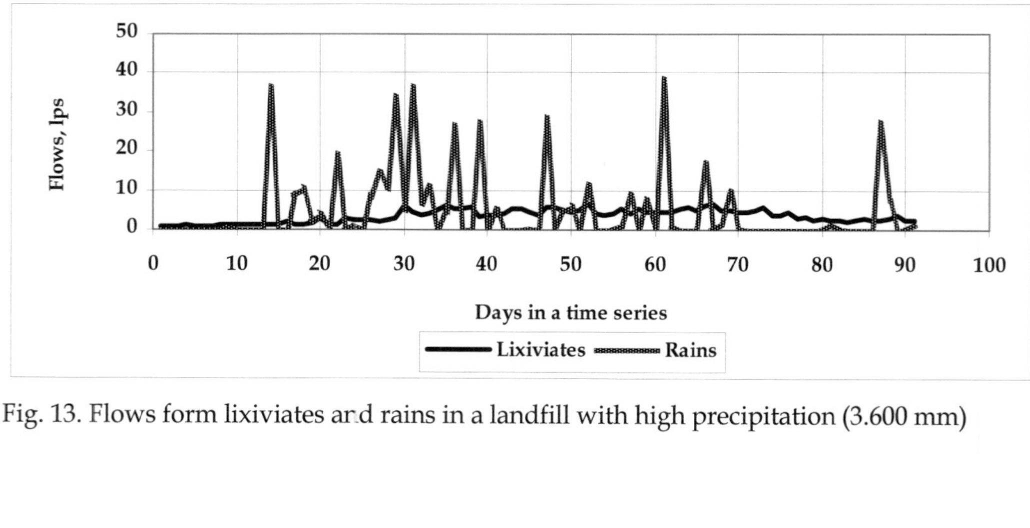

Fig. 13. Flows form lixiviates and rains in a landfill with high precipitation (3.600 mm)

rain was distributed on the land fill site area and the resulting flow is compared to the actual landfill lixiviates flow. Its is clear that rains will increase lixiviates and complicate the water treatment system for them, as water treatment plants have investments cost related to flow to be treated.
It is estimated that the flow of lixiviates coming form the process is caused basically by the organic materials deposited in the last 2,5 years in the landfill site. The contribution factor was estimated as 0.0041 litres per second (lps) per each 1.000 of organic ton of this nature. The flow of additional lixiviates coming from the effect of the rain is estimated with the following expression (where x is monthly rains in mm):

$$\text{Additional lixiviate flow from rains, lps=}$$

$$\text{Area that allows drainage (has)}/5 * (0{,}00000832x^2 + 0{,}00430x + 0{,}54) \quad (1)$$

Lixiviates have to be treated. The investment in the treatment plant is important and may be in the range of 0.20 to 0.40 US million $ per each litre per second of lixiviates. The treatment cost as such could be in the range of 0.50 to 1.20 US $ per m3 of lixiviate.

Biogases generated

Biogases are generated also because of the organic nature of waste, under the effect of bacterial action. Biogas will tend to be a mixture of CO_2, CH_4 and humidity.
Biogas is a rich resource and can be used to generate electricity or heat or even hydrogen. However, this will depend of how strategic the decision of placing the landfill site was. If the site is far from the practical use of these applications, then the energy recovery costs will be too high and the gas will have to be burned to avoid green gases emissions without real gain. This sadly will happen because people tend to reject landfill sites, so they are placed far from the cities of industries. An alternative is to conceive the site as an industrial integrated unit able to generate added value in many ways, all of which will require energy and electricity. The recycling facilities that transform plastic in final products, for example, are large consumers of energy and can be close to the land fill site.

Typical biogas composition (% volumetric wet)	
CH_4	47,7
CO_2	29,6
H_2S	2,4
H_2O	20,0

Table 8. Typical landfill site biogas composition

Biogas annual emissions from organic deposits can be estimated using the expression (based on John Pecey 1975)

$$\text{Emissions (Nm3/min/ton organic)} = 0{,}0000986\ x^{-0{,}756} \quad (2)$$

Where x is the year counted from the initial placing of the organic material in the site. The behaviour is exponential and shows decreasing amounts of emissions. After 20 years the total emitted will be around 0.319 Nm3 per kg of organic waste. The high heat power contained in the biogas will around 4.240 Kcal/Nm3 of wet gas. About 25 to 35 % of this energy can be converted to electricity.

With this information estimates can be made of the biogas emissions and of the potential they have to generate electricity or heat.
To gather the biogas, wells should be placed and also a pipeline network. A rule of thumb is to place each one to collect 1.600 m2 of landfill terrain. Each well will handle around 60 Nm3 of wet biogas per hour, obviously depending on the real design of the landfill.

Incineration

This is a real possibility, because the waste contains important amounts of energy. Plastics, paper, cardboard, wood, textiles contain net heat power. Organic materials, being wet, do not contribute as much, although they also have energy potential when dried. Incineration has the advantage of reducing significantly the volume of waste to be disposed, as much of the mass become gases (mostly CO_2 and water) and only the ashes stay solid, which will be less than 10 % of the initial matter. Also, it can be done in a compact facility, within the city limits, fulfilling all the environmental norms when well designed and operated. For these reasons many cities in Europe, Japan and USA have incineration facilities. However incineration is quite questioned nowadays because it destroys valuable materials such as plastics and contributes to green house effects.
Any strategic analysis should include incineration and an as complete as possible impact analysis should see the different sides of the situation.

7. A comparison of possible alternatives for solid waste handling

As an example of strategic analysis, a situation is presented, in which four different alternatives are studied. It applies to a city with three million people, situated in a developing country, generating some 2.000 tons per day of waste, that are taken mostly to a landfill site with only small percentages being subjected to recovery or recycling..

Alternative 1, the present situation

In this, there are no programs of recycling or separation in the source, undertaken directly by the organization in charge of waste handling, although some recycling and separation are under way done by low paid recyclers and some well organized cooperatives. At the time, a modest organic waste recovery program for elaborating compost was in place, which did not work. This alternative does not include a plant for treating lixiviates nor any systems to take advantage of the energy gases.

Alternative 2, an improvement on the current situation

It is an alternative that is thought to be reachable, but implies concerted efforts and intelligence of all the involved sectors, much conscience in the city, leadership in the organization and citizen commitment. It includes a plant of treatment for lixiviates, a thermal plant to take advantage of the energy of biogases, a rotary incinerator for special waste, increasing separation in the source and of recycling programs, with strong educative components and a more intense program of treatment of the organic wastes towards composting.

Alternative 3, one more aggressive approach to recycling and separation

It includes more demanding goals for recycling, treatment of organic and separation in the source, simultaneously with treatment of lixiviates and also using a thermal plant for recovering biogas energy and a rotary incinerator.

Alternative 4, aggressive education and source management of waste

It is a situation in which the emphasis is put mostly in recycling, treatment of organic waste and separation in the source and a lixiviate treatment plant is included. It does not consider thermal plant or incineration to take advantage of the biogas energy, which is simply burned. Investment in education and publicity is high.

Results of the model

Alternative	1	2	3	4
Situation	Actual	More recycling and biogas recovering	High recycling and high biogas recovery	Very high recycling and not biogas recovery
Waste recycling and separation	No	Medium	High	Very high
Lixiviate plant	No	Yes	Yes	Yes
Organic waste treatment	Very low	Medium	High	Very high
Thermal Plant	No	Yes	Yes	No
Incineration	No	Yes	Yes	No
Waste generated and treated in 20 years (million tons)				
Total generated waste	19,62	19,62	19,62	19,62
Organic treated	0,27	2,02	3,91	4,85
Recycled or separated		1,90	3,53	4,23
Incinerated		0,38	0,38	
To be taken to the landfill site	19,35	15,32	11,80	10,54
Other operation parameters				
Open area in the landfill site (hectares) in 20 years	154	122	95	85
Biogas energy flow (year 20), million Kcal/hr	58	42	25	21
Number of trips to the site (thousands) in 20 years	0	1.962	1.771	1.608
Directly Employed persons in the programs	875	1.072	1.203	1.177
Relative economics, % savings against actual situation		7,17	6,37	1,94

Table 9. General results for the four alternatives for a 20 year operation of the programs

It was observed that the alternatives 2, 3 and 4, that imply programs of separation in the source and treating in some way the waste, are more attractive from the economic point of view that the present situation. In spite of the apparently high investments, the alternatives that contemplate use of biogas are more attractive than those than do not use it. It is clear that this is not really an economical problem, but a problem of beliefs and attitudes.

It is important to note that the proposed alternatives, besides being more economical, show also a favourable impact on direct and indirect employment, which is very important for a city like the one studied, subject to difficult unemployment and social problems. These cities have enormous potentials for growth and a real need for improving the quality of life for the people. It should focus correctly its solid waste problems. It has been shown that alternatives that attack the problem from an integral point of view are more attractive than the current situation. It is necessary that a city of this importance count with integral planning for handling this type of problems.

8. References

Posada, E. (2009). The role of poetry and images in creating recycling and resources saving awareness, *R'09 Twin World Congress on Recycling 2009,* Ecotopia Science Institute University of Nagoya, Nagoya, Japan

Posada, E. (2008). A contribution to the strategic analysis of alternatives for waste management in Valle de Aburrá Area, Medellín-Colombia , *R'08 Global Symposium on Recycling, Waste Treatment and Clean Technology REWAS 2008,* TMS, Cancún, México

Posada, E. (2005). Creativity applied to recovery and recycling, *R'05 World Congress on Recycling 2005,* EMPA, Beijing, China

3

Professionalism Pays: Industry Associations and Continuing Professional Development for the Waste Management Sector

Dr Georgina Davis
Griffith University
Australia

1. Introduction

Some key premises for understanding the development of occupational competence and attachment are how individuals come to identify with that occupation and what support provisions (including qualifications and continuing professional development) are made accessible in developing both their occupational capacities and sense of self as a practitioner or professional within the sector. Together, these comprise key bases for the formation of occupational competence and the premises for individuals association with and learning for particular occupations, such as waste management. Yet, such bases are not consistently afforded across occupations, with some enjoying greater esteem and development opportunities than others, and also differentiations in the kinds of educational provisions and sector support being afforded to offer engagement and learning.

This chapter forms a review which critically evaluates the role of professional membership bodies and discusses if some of the continuing professional development (CPD) approaches and rationale used in more established occupations can be successfully applied to other emerging professions/industries (who are emerging as new professions or are varied in their stages of development across different countries). It also specifically seeks to identify the benefits and role of professional membership for individuals working within waste management functions and how these vary between countries. The consideration here is to assist in the identification of both current and future training and educational requirements, including the kinds of certifications (i.e. qualifications) required to support the standing and development of this growing sector and providing a skilled and professionally recognised work force. This consideration and elaboration should assist understanding bases by which an emerging sector of employment, such as waste management, can engage with those who work in the sector and be supported by institutional arrangements. The goal is to appraise how such arrangements could be strengthened and demonstrated through a brief overview of the current standing of professional membership of waste management professionals within Australia, the United Kingdom (UK) and within an international context (the International Solid Waste Association who provide an 'umbrella' organisation for other waste management organisations and institutions).

The review looks at the role and benefits afforded by professional intuitions and introduces the mechanisms for recognition of professional waste managers. The chapter concludes by discussing the development opportunities for a sector to self-regulate through professional standards and the improvement of educational and training support systems; and where the mandatory regulation may provide both opportunity and constraint.

2. The role of professional industry associations

A professional industry association can be described as an organisation which represents the interests and development of a particular discipline or profession, and those individuals working or studying within that discipline or profession. Whilst there are numerous professional industry associations worldwide representing an extensive range of job functions and activities, these associations vary in the levels of service and classes of membership offered to their members and the profession/industry that they represent. In the most sophisticated cases, these industry bodies facilitate personal development and advance professional recognition for their relative disciplines, job functions and members; and actively engage the profession and other stakeholders through a variety of activities. It is essential that a professional institution is perceived as representing its members interests and will support and enhance the professionalism of the industry and the professional development of its members. This will attract members and the support of industry organisations who may themselves wish to become members (where possible) or who may provide support to individual employees who wish to become members. This support may include time off work to attend professional events including conferences, specialised training, volunteer roles within the industry association and branch meetings; and in some cases, the payment of annual membership fees.

Professional industry associations can provide a critical role in developing occupational and professional standards for their respective disciplines and associated job functions through close collaboration with industry and its membership, so as to ensure that any qualifications and training meets the current skills need of the profession and its employers (Davis, 2009). As the pre-eminent source of labour market information for the profession and raising the profile of the profession, including ensuring individuals are aware of the opportunities for development and employers recognise the benefits of investing in the skills of their employees (London Economics, 2008). Industry bodies also may have a responsibility to promote and educate their professions and undertake associated activities such as research development, knowledge dissemination, lobbying, marketing and policy development (London Economics, 2008). It is necessary that an industry body offers a range of education and training activities (not just those formally accredited and those leading to professional qualification). For example, one-day seminars, technical symposiums, conferences and networking events. This will increase the appeal of affiliation and membership to a broader audience and, more importantly, offer greater development opportunity to a sector/profession and those individuals working within it as a whole.

Professional industry bodies and institutions tend to be governed by their members and, as such, reflect the requirements of the industry as perceived through its members, be those individuals, corporate organisations or a mixture of both. Professional industry associations also have a code of conduct by which its members are bound. Members (individuals or organisations) who act in violation of the code (where a complaint is investigated and upheld) may be subject to disciplinary actions which usually result in the removal of membership from the institute. In this way, these associations are held to be self-regulated to a degree and it is this autonomy that is guarded and often the basis for professional autonomy. Yet, there are limits to this autonomy and occupations that are highly regulated may be subject to external regulation and monitoring. So, just as the certification and monitoring of commercial pilots is managed by government departments, despite there being professional associations, the handling of hazardous waste (e.g. radioactive, healthcare) is subject to external regulation through national and international legislation and formal agreements.

2.1 Identity and membership

Individuals are likely to join groups or actively seek memberships for a variety of reasons (Gallagher et al., 1997):

- as a mechanism for fulfilling personal interaction and affiliation needs;
- as a means for fulfilling status and self-esteem needs;
- to assist individuals in establishing a self-concept (that is, a way of defining their own identity); and
- as a mechanism for achieving power and goal-achievement needs (through a collective power and status).

These reasons emphasise benefits for the individual on a personal and professional level. Other reasons for joining such associations may include access to particular jobs and functions (where professional status is a requisite) and the associated salary and title. The collection of group memberships that an individual has and the attribution of unity across these different memberships can strongly influence one's identity (Pullen et al., 2007).

Swank (1987) also observed the need for individuals to be associated with an organisation that provided them with 'professional stature'. Bennett (2000) undertook a survey of individuals and organisations to determine the logic of sectoral business associations in the UK. The survey findings indicated that the role of accreditation to individuals was the primary driver for joining. The study went on to conclude that industry associations with larger numbers of individuals and/or small businesses as members placed a greater importance on collective activities with the main priority being accreditation for individuals.

However, beyond forming an occupational identity, there are issues associated with maintaining or refining that identity. For instance, Allen (1963) proposed that to achieve a genuinely professional attitude and therefore professionalism, it is necessary for an individual to be aware of the scope of their sector, keep up-to-date with developments, participate in associated activities (both social and non-social events) and encourage and support progress within the profession. Allen (1963) also stated that professional (scientific) organisations provide many advantages including:

- providing identification of and interaction with other professionals;
- opportunities of attending symposia, conferences and other networking and learning events;
- opportunities afforded through professional contacts;
- providing journals, newsletters and other industry related information;
- providing a media for publication;
- discounts for access to events and journal publications.

It follows that the professional identity of an individual is essential and invaluable to the status of a discipline/sector and highlights the importance of professional identity as being akin to personal identity and that professional identity can eventually parallel professional recognition (Allen, 1963). So, there is reciprocity here as the development of the sector requires development of the individuals and organisations acting within it. A Waste Management Association of Australia (WMAA) Survey conducted in 2009, (Inside Waste, 2009) determined a range of factors for individuals to join the association. Whilst the largest category (24%) reported that membership was to ensure that they kept abreast of the latest technology and trends, 20% reported networking opportunities; 15% to foster industry professionalism; 13% to receive newsletters and publications; 10% to access education and training; 5% stated discounts to WMAA events, and finally 2% specified 'other' reasons. The WMAA survey findings therefore are consistent with the literature in terms of what they offer to individuals, including factors for joining and also the advantages they provide (see Gallagher et al., 1997 and Allen, 1963). The survey results also highlight the importance that WMAA members associate with maintaining current and up-to-date knowledge of the sector and ensuring professionalism throughout the sector.

2.2 General benefits

Previous studies have determined the benefits (economic and/or personal) of attaining higher qualifications (Walker & Zhu, 2006; Royal Society of Chemistry, 2005). However, there are few well documented studies detailing the benefits associated with professional membership, perhaps because many of the associated benefits, such as networking opportunities and life-long learning through CPD, are more difficult to quantify or still lack acknowledgement of their importance by some employers.

In December 2008, the Consultative Committee for Professional Management Organisations (CCPMO), an industry advisory committee, released a report outlining the benefits of professional membership to industry, the career of individuals and the wider economy as they had determined this area had not been addressed in previous studies. The CCPMO represent eight UK professional bodies (Institute of Credit Management; Chartered Institute of Logistics and Transport; Chartered Management Institute; Chartered Institute of Marketing; Chartered Institute of Personnel and Development; Chartered Institute of Management Accountants; Chartered Institute of Purchasing and Supply; Institute of Chartered Secretaries and Administrators) with over a combined 1,500 employees and 560,000 students. Annual turnover for these organisations range between GB£2-43 million with a clear relationship between the annual turnover, number of employees and the total number of members (including students). The CCPMO also award over 50,000 qualifications per annum ranging from entry level qualifications to post-graduate level with these opportunities extending to both UK and international applicants (London Economics, 2008). The courses leading to these qualifications are developed in close collaboration with industry and other key stakeholders and are therefore described as "demand-led" and are offered through a variety of learning media including accredited education providers; in addition to accredited university courses and modules. The members of the CCPMO either operate directly as, or contain a Qualifications and Curriculum Authority (QCA) recognised awarding body and, as such, have to comply in accordance with UK national standards relating to quality assurance and regulation [QCA recognises and regulates awarding bodies and their qualifications in the UK in order to maintain the standard of the national qualifications framework (UKCES (2008)].

There are seemingly direct benefits to members of professional associations in terms of salary and the worthiness of the kinds of employment they secure. The CCPMO research concluded that individuals holding "*professional qualifications and membership of a professional institution are estimated to achieve both higher earning and be more likely to be employed...in comparison to individuals with no professional qualifications*" (London Economics, 2008: *pviii* and p32). The financial data indicating a lifetime economic benefit of both professional qualifications and membership as being approximately £152,000 (comprised of £81,000 from professional qualifications and £71,000 from professional membership). Such a financial benefit to the individual clearly provides a strong incentive for undertaking professional qualifications and pursuing membership which, it could be argued, ensures life-long learning through CPD is undertaken whilst the individual remains in employment. Additionally, within this sector many salary scales need to be supported by a professional qualification (Heynes, 1994), not unlike Chartered Waste Managers in the UK who can apply for a wider range of positions (mainly management roles) and can additionally expect a higher salary of around £3-5,000 more than their unrecognised counterparts (LearnDirect, n.d). Harvey et al., (1994) also determined that salary was an indicator of perceived professional standing and success.

Although the report (London Economics, 2008) clearly represents a strong marketing tool for membership of these institutes, economic and employment opportunities associated with professional membership have been noted elsewhere (Harvey, et al., 1994; Davis & Read 2006). Additionally, such activities must represent a broader benefit to the economy through increased income tax payments but also the increasing professionalism of a sector resulting in improved practices and often accompanied by tightening regulation. The members of the CCPMO and other professional industry bodies (such as the UK's Chartered Institution of Waste Management and the Chartered Institution of Water and Environmental Management) are also self-funding, requiring no financial support from government to develop and deliver their training and CPD activities. The profitability associated with the provision of training and professional qualifications in some industries has led to an extensive network of accredited training providers and development of a flexible range of study materials which, in turn, has increased the flexibility associated with participating in further personal development. Thus, allowing many to undertake further study in collaboration with full-time employment and more importantly, where learning compliments job function, receive funding from their employer.

A UK study (London Economics, 2008) detailing other qualifications held by individuals with professional qualifications and members of professional institutes determined that approximately 10% of the UK's working population hold a professional qualification whilst approximately 2% are members of a professional institute. Additionally, the qualifications held by those individuals with professional qualifications and/or professional membership vary significantly from the rest of the working population. For instance 33% of individuals with professional qualifications have undergraduate degrees, whilst this rises to 57% for individuals with professional memberships. When the data was further broken down individuals with professional membership also undertake a range of other qualifications and studies throughout their career indicating a process of life-long learning.

3. Current recognition of waste management professionals

This section seeks to highlight the differences in professional standing and development opportunities between individuals working in the waste management sector across two countries, the United Kingdom (UK) and Australia. Despite similarities between these two countries relating to industry trends and synergies such as increasing consolidation of the industry (IBISWorld, 2007) and the prevalence of the same international organisations; the recognition and support of waste management professionals is diverse.

The Waste Management Association Australia (WMAA) is currently the only specific waste sector industry body, although other institutions such as the Environmental Law Society and the Environmental Institute of Australia and New Zealand (EIANZ), represent some individuals working within the waste management sector. The WMAA is a 'young' organisation when compared to the UK's Chartered Institution of Wastes Management or the International Solid Waste Association (ISWA), with over 100 years and 35 years respectively of constitution and membership. Membership of WMAA is based on subscription and there are no formal training or educational standards which exist or regulate/certify levels of membership. WMAA offers a wide range of learning opportunities for both its members and other interested bodies through seminars, breakfast meetings and conferences. However, there is no mandatory Continuing Professional Development (CPD) specified for its members or Structured Educational Training (SET) for its graduate members. Therefore, the basis for membership, categories of membership and structured professional education arrangements within the WMAA are under-developed (Davis, 2008).

The professional recognition of industry practitioners is an essential part of developing the waste/resource management sector. In order for structured professional development opportunities and professional recognition to be implemented, it is essential that the functions/roles within the industry (both present and future) are accurately mapped. From which, suitable training and education programmes can then be identified and developed; and membership classes proposed and entry requirements to those classes determined. As a result of the emergence of the waste management sector and level of sector development in Australia, there are currently limited education and training programmes both within, and for, the Australian waste management sector, particularly when compared to other countries such as the US and UK (Davis, 2008). The formal recognition of the waste management profession and the professionals operating within the industry has in the UK for example, been a significant driver for the development of training and education programmes (i.e. professional membership schemes including Chartered status). Many UK universities include waste management modules within a range of degrees including environmental studies and engineering. With little demand from students and industry in Australia, and the lack of recognition of the importance of the sector its omission from units/modules in key disciplines such as engineering is set to continue. Additionally, in the UK there is an extensive range of stand-alone courses ranging from MBAs to post-grad and undergrad degree programmes specifically in waste management (Davis, 2008). The standing of qualifications and the level of qualification is proposed to be an indicator of the standing of the occupation.

The UK's Chartered Institution of Wastes Management (CIWM) specifically represent individuals as opposed to corporate entities. The CIWM has over 7000 individual members whilst WMAA has under 600 individual members (including students). However, unlike CIWM who only offers individual memberships, WMAA has over 150 organizational memberships (including state corporations, small businesses and Local Governments).

There is also a distinct difference in the development and complexity of the membership grades and eligibility criteria between WMAA and CIWM. The diverse range of academic and competency based qualifications in the UK has allowed the CIWM to purposely break down and segregate different levels of educational and professional achievements attained by its members facilitating a diverse number of membership grades, but also clearly articulating and allowing movement (development) between them. Additionally, the CIWM offers a further professional designation (Chartered Waste Manager).

4. Identifying and recording competencies for professional membership

Professional institutions set criteria for key competencies for individuals to achieve against classifications of membership. The CIWM provides an overarching set of its competencies designed for full Corporate Membership and Chartered Waste Manager status (Table 1). Table 1 shows the five broad key competency areas and the types of activities within each which must be adequately fulfilled to be eligible for Chartered status. These competencies reflect the organizations role in lifting the professionalism with the sector through setting professional standards, essentially providing a benchmark of minimum competencies required to fulfill the roles and job functions reasonably expected as a Chartered Waste Manager in the UK.

	Key Competencies: CIWM Chartered Waste Manager
1	Knowledge and understanding of the wastes management industry, including: • Current and impending legislation; • Waste Strategy; • Hot topics and current affairs; • Structure of the Industry.
2	Ability to analyse and evaluate problems and develop practical solutions: • By providing examples from their own working experiences as to how these competencies have been met; • Demonstrating creativity, innovation and motivation.
3	Leadership in the management of waste, giving examples of: • Team management; • Project management; • Motivation; • Monitoring and support; • Promotion of sustainable waste management.
4	Effective interpersonal skills providing examples of: • Written communication • Presentations; • Engaging information and providing advice; • Chairing meetings and committee representation; • Technical publications and reports.
5	A personal commitment to professional standards recognizing the obligations to society, the profession and the environment, providing examples of: • Promotion of sustainable waste management; • Engaging with key stakeholders; • Working within legislative and regulatory timeframes; • Personal development and training.

Table 1. Competencies for CIWM full Corporate Membership and Chartered Waste Manager status (CIWM, 2008; www.ciwm.co.uk).

In 2004, the International Solid Waste Association (ISWA) developed the International Waste Manager (IWM) qualification, launching the scheme mid-2005 and with their first applications being received in early 2006. The IWM certification is awarded at three different levels, intermediate, advanced and international depending on the applicant's ability to meet the competency criteria specified in Table 2. Overall, the competencies are close to those specified by the CIWM such as making a commitment to sustainable development, understanding legislative commitments and requirements, and a range of management and administrative functions. Additionally, ISWA requires a commitment to their Code of Ethics "*To take steps to minimize environmental harm; use skills and experience in waste management to serve the needs of the environment for responsible environmental behavior; not to encourage conduct involving dishonesty, fraud, deceit or misrepresentation or discrimination; and commit to maintaining personal professional competence and strive to maintain integrity and competence of the profession*" (CIWM, 2009: page 20-21). Such commitments to codes of conduct or codes of ethics being integral to all institutions.

	Key Competencies: ISWA International Waste Manager
1	The applicant must be able to demonstrate knowledge and understanding attributes of: • Sustainable waste management principles; • General management and administrative procedures; • Thorough understanding of legislation and regulations relevant to country and region where operating; • Identify and explain short, medium and longer term environmental threats and opportunities related to the sustainable management of waste.
2	The applicant must demonstrate the following competencies: • Develop and communicate waste management issues to a wide range of audiences; • Demonstrate a willingness and persistence in addressing abnormal waste management issues; • Possess a high level of leadership and motivational skills to ensure that good waste management practice is effectively communicated and integrated into decisions and actions; • Identify, engage and respond to stakeholders; • Develop effective means with which to liaise and advise others.
3	The applicants must be able to demonstrate their engagement by: • Understanding a range of global threats and their importance to the waste management industry; • Identifying solutions to environmental improvement and mitigation and recognizing their dynamic nature; • Recognizing the interdisciplinary nature of waste management issues; • Putting environmental issues into their working context; • Demonstrating that they approach work in a competent manner and work towards and secure change and improvements; • Identifying measures to ensure that individuals and organizations are accountable and understand their responsibilities for both environmental damage and improvement.

Table 2. Competencies for ISWA International Waste Manager status (CIWM, 2009; and www.iswa.org)

The one defining tool for awarding and retaining professional status across the various disciplines and institutions is continuing professional development (CPD). Membership grades which require a structured CPD framework to be completed on an annual basis by its full members or structured entry to a graduate membership grade need to be fully supported through the professional body awarding that membership. CPD is an essential requirement to retain Chartered membership grade in any institution. For example, CIWM has developed strong frameworks for both CPD and Structured Educational Training (SET) specifically for its graduate members. Whilst such a framework could be viewed as a mechanism for members to take control of their own development and professional development needs (Noon, 1994); they also provide a mechanism for "taking stock" (Wilson & Halpin, 2006) as can be clearly identified in Table 3 which shows the electronic 'CPD Recording Spreadsheet' developed by the CIWM for completion by all individuals holding a professional membership grade. Although the table headings are 'blunt' and the area for reflection is weak, the sheet provides a visible record which can easily be used and later retrieved for the purposes of CPD verification by the institution and for personal development by the individual such as for the construction of a resume.
CPD is perceived as an essential component of entering and remaining within a profession and can influence the construction of professional identity (Tang and Choi, 2009). The rationale for CPD has undergone a paradigm shift over recent decades. Jackson (1968) proposed that CPD was essentially a 'deficit model' used to fulfil gaps in the knowledge and skills of individuals; as opposed to an aspirational model which implies that improvement is being undertaken. Hargreaves (1994), proposed the post-technocratic model of professional development in which lifelong professional learning occurs which is regularly assessed and which is also reconciled against the organisations needs. Days and Sachs (2004) however, propose that the technocratic model is flawed. As individuals move through different stages of their careers, their needs change accordingly and these may be inherently different from the needs of their employer/organisation (thus creating another area of potential conflict).

Date	Activity	No. of Hrs	Reason	What Learnt	How will I use it/further action	Example

Table 3. CIWM CPD Log.

In addressing this difficulty, the UK's CIWM SET model covers core skills required unilaterally across the profession but also allows individuals to undertake role and discipline specific training and development activities. Additionally, CIWM courses (most of which are accredited and often operated in collaboration with external parties) cover a range of technical areas (Davis, 2008). In 2009, the CIWM further expanded its support for core skills across the disciplines and entered collaboration with the Open University to offer individual non-accredited modules in subjects from financial management to team work. This is essential for such a diverse profession which embraces many disciplines and activities. It has been argued (Reid & Brown, 1996 as cited in Wilson & Halpin, 2006) that the growth of core, highly transferable and cross-disciplinary skills is damaging to the concept of profession and affords movement of key professionals between sectors, to the detriment of poorer salaried or less prestigious professions. The perceived levels of professionalism within an Institution and commonly held standards between all Chartered Institutions also facilitates the movement of professionals between sectors, allowing them to gain professional membership to other institutions.

5. Self-regulation and the role of industry associations

Bennett (2000) highlighted the ability of professional bodies and other associations in the self-regulation of industry sectors, helping to improve standards and competitiveness of both organisations and individuals. However, Bennett (2000) also highlighted the issues associated with an industry body trying to be as all-encompassing as possible which leads to professional standards being dumbed down and compromised down to the lowest common denominator.

Higher expectations and requirements for an industry require suitably well qualified individuals who possess the right knowledge and skills upon entry to the profession but also throughout their careers. Professionals historically have been neutrally portrayed as 'restricted' or 'extended'. However, the change for professionals to now be 'compliant' is coupled with the expectation for many professions that CPD is now an expectation as opposed to an option (Hoyle, 1980) as is the membership of appropriate institutions and professional bodies. This is now certainly true for some engineering professions in Europe, where only Chartered Engineers can legally 'sign off' key documents, and where Chartered status now requires a minimum of a master degree and numerous core competencies to be fulfilled during academic studies (Davis, 2006). Doney (1998) and Noon (1994) as cited by Wilson and Halpin (2006), both positively discuss the role of mandatory CPD for a sector with regards to raising the 'standing of its professionals' and 'professional self-image'.

Government policies and interventions can be a key driver for CPD and the development of CPD systems. The UK's waste management sector for example, has seen increased structuring and formalisation of CPD for its practitioners in order to drive accountability and performativity. Although this has no doubt raised the perceived professionalism of the industry and has even resulted in increasing salaries for those holding professional memberships and qualifications (Davis & Read, 2007); the driver for this development in the UK has been largely due to legislative requirements and agendas rather than for the sake of the development of the profession or the individuals within it. Historically, personnel working within the UK waste management industry tended to have minimal educational qualifications and acquired initial practical skills through 'on the job training' within the sector. However, the Environmental Protection Act 1990 and the Waste Management Licensing Regulations 1994 introduced the concept of 'technical competence' for waste management personnel in positions of responsibility. The aim of these regulations was to assess personnel competence to a recognized occupational standard in order to ensure more sustainable management of wastes.

The Waste Management Regulations (as amended) 1994, stated that all managers of facilities covered by a waste management licence needed to demonstration their technical competence and, therefore, require a Certificate of Technical Competence (COTC). The COTC was obtained through the vocational qualification schemes, which were achieved through the assessment of an individual's competence to do the job against national standards. The required vocational qualification (VQ) depended on, and related to, the particular area in which the individual worked. Upon completion of the VQ, an individual then applied for the corresponding COTC. VQ's are broken down into National Vocational Qualifications (England and Wales) and Scottish Vocational Qualifications (Scotland) and are offered at Levels 3 and 4 for the mandatory COTC. In addition to the mandatory COTC, there are a wide range of National Vocational Qualifications (NVQ) and Scottish Vocational Qualifications (SVQ) that are non-mandatory and are designed to improve the skills of the workforce. These qualifications cover a range of waste management sectors (collection; treatment; landfill; recycling and street cleansing) at Levels 1- 5. The N/SVQ's all focus on the practical ability of an individual to perform their role and, as such, are assessed through direct observation and proof of performance.

Increasing regulation of the waste management sector in the UK has seen a rise in the number of professionals 'testing' their professionalism and drawing on professional certifications to prove competence and validate their findings/work. Certainly, the credentials of an individual acting in an 'expert witness' capacity are more readily accepted and proven where certification is held.

Mandatory national certification for professionals may also be considered. Under such a programme applicants would need to meet all qualifying criteria. Such an approach would assist industry regulators by allowing them to use certification as a condition of licencing both environmental professionals (for example, consultants providing reports) and the operational and management staff located on licenced waste management sites, as is currently the case in the UK. It could also act as a marketing tool for some professionals who can use their certification as a means of distinguishing themselves from less qualified individuals. Wilson and Halpin (2006), make the distinction between training (including CPD) that meets the needs of the sector as opposed to the profession. There is concern that where mandatory qualification is required, training opportunities are focused on attaining the qualification and ensuring ongoing compliance, as opposed to identifying and undertaking CPD which is beneficial to the individual and the organization; and during times of limited budgets and training opportunities, it will be less likely that any CPD activities extend beyond compliance.

Based on the requirements and competencies for the professional membership classes required by the CIWM and ISWA (Tables 1 and 2), it is therefore proposed that there are three conceptions of knowledge associated with the development and learning by waste management professionals (as adapted from Cochrane-Smith and Lytle, 1999: 'conceptions of knowledge associated with teachers learning and development'):-

- Cognitive knowledge for industrial practice: formal knowledge generated through research and industry collaborations, commonly documented in the peer-review and trade press media. Includes government funded research and formalised initiatives. Particularly applicable to the dissemination of new waste management technologies.
- Knowledge of practice: critical examination of personal performance against job description/role. Knowledge gained through experience and undertaking the role. Including the recognition of broader business responsibilities, and social and environmental concepts such as sustainability/sustainable waste management.
- Self-knowledge: practical knowledge gained through 'on-the job' experience, evaluation of previously attained qualifications; and education and training opportunities completed. Including those from development and career objectives. This would require reflection of personal issues and values, and all new knowledge (from all sources).

Such a framework of knowledge provides opportunities for the professional development of an individual and the sector. Any expansion to provide additional membership classes for the Australian waste management sector must accept the forms of knowledge held by waste management professions and fairly acknowledge achievements.

6. Conclusions

There is differentiation across all countries with regard to the certification and opportunities afforded to professional indivduals working within the waste management sector. Although ISWA has provided an internationally recognised grade of membership, the lack of perceived prestige of the sector in some countries diminishes the standing of this award. There are opportunities for the further development of both professional memberships/ associations and support mechanisms for this valuable industry in many countries.

Any professional certification or qualification, particularly those leading to professional membership grades needs to consider 'open access' (that is access to all members of the profession regardless of academic achievement or time spent in industry). In order to achieve open access, a range of professional qualifications across all levels is required; and the development of membership grades to reflect those qualifications so that access to membership and qualifications is appropriate to various entry routes. There is a danger where membership grades are based purely on professional qualifications (including degree level attainment); that the membership body is viewed as elitist or 'closed' to certain groups of individuals. Given the diversity of professions and skills encompassed by waste management and the considerable industry knowledge and expertise held by many practitioners, access to membership grades also needs to be based on industry experience and, as such, the development of a suitable mechanism for recording, recognising and accrediting appropriate industry knowledge and experience is essential. This then allows all learners and industry personnel the opportunity for accessing professional membership and the benefits associated with it even if they do not hold formally recognised qualifications. This discretion is essential for multi-disciplinary sectors. For such a diverse sector as waste management, there is no rationalization for the formation of steep barriers for active membership based on particular disciplines or degree courses. However, membership based on more than purely fees paid and compliance to a Code of Conduct are necessary for professional development of the sector. Given the diversity of the disciplines across the sector, it would be desirable to allow all suitable degrees and other qualifications to be eligible for a certified level.

Who decides the requirements for being classed as a 'waste management professional' and what the bases are for that determination will be key to the process. This question is fundamental in the design of standards for certification/accreditation and measuring competency and attainments against those criteria. For example, in the first instance a judgement on the level of educational attainment must be made; how many years experience working in the industry is required; types of work and roles in certain disciplines/sectors also need to be decided. If poorly consulted or conceived these decisions can negatively influence membership. Merhr et al., (2002) acknowledged that for example, the US Wildlife Society has in its past been too linear with regards to assessing applications for professional certification with applicants being denied certification based on the degree they did or did not do or their college transcript; with the Committee on Professional Standards stating that the "*academic and experience requirements as applied must continue to provide an indispensible measuring stick for nearly all applications….. even though there is no justification for the erection of impassable barriers to active membership based solely upon such concrete requirements as courses taken, degrees received or positions held*". More recently, the Committee was provided with more discretion so that applicants only had to 'satisfy the intent' of the requirements. Maehr et al., (2002) also identified a number of perceived barriers for individuals applying for certification including:-

- The application process is too time consuming;
- Certification is too expensive;
- Any changes to the criteria for certification of individuals can lead to a perception of an uneven playing field;
- Perceptions that poor candidates have obtained certification whilst quality practitioners are refused;
- Certification is unlikely to provide any benefits, particularly for those already functioning at a professional level within an industry.

In instances where there is industry resistance to adopt an accreditation system or where even certified individuals resist or fail to undertake ongoing professional development and to demonstrate that professional development, the sector may experience an increase in the number of practitioners who are poorly trained (Maehr et al., 2002).

The process of accreditation of university and other educational courses has long been recognised as a way of formalising a sector and imposing quality. However, a professional institution also needs to determine a robust methodology agreed with academia for the ongoing monitoring, regulation and development of courses. This can require an adjustment of views, particularly in some institutions where the control of curricula is left to the discretion of individual academics who teach their area of expertise as opposed to a wider and broader knowledge set (Davis & Read, 2007). Whilst this diversity in taught curricula for the same sector could be seen as a benefit as it recognises this diversity and educates individuals to undertake different functions and allows academics flexibility to teach to their agendas, it may also be seen as a disadvantage indicating that educators, employers and professional bodies have failed to develop and/or communicate common standards for educational programmes. The promotion of professional bodies may also be weak within some higher education environments. It is therefore proposed that the accreditation of a course and/or certification of professionals would assist in the promotion of quality and professionalism within higher education and training courses.

Due to the diversity of the profession and its interdisciplinary nature, it will be harder to define a coherent core curriculum. The depth of knowledge required for specialisation in some areas versus the breadth of the sector and issues impacting the sector would need to be carefully considered and balanced against any requirements for professional membership grades/certification. Several attempts have been made by various organisations across different countries to initiate the development of a waste management curriculum. One such attempt was by the National Recycling Coalition, US, who assembled an expert committee to devise an integrated waste management curriculum (Conn, 1993). The committee first established a set of four objectives for a curriculum (Conn, 1993) which comprised: (i) the education of specialists in integrated waste management (IWM); (ii) to provide training in IWM skills; (iii) to provide non-specialists with a limited knowledge of IWM; and (iv) to contribute to the development of environmental literacy amongst students generally. These four objectives provided a rational approach to meeting industry and academic requirements, creating a starting point, which identified sector needs before attempting a solution. Clearly, a single curriculum would evidently not meet all of these objectives as there is an individual need for a unique and specific curriculum and/or approach to meet each objective. This is the raison d'être applied to, and driving ongoing research, where it is accepted that the curriculum for under-graduate students will be different from post-graduate students which, in turn, will be different from courses designed to meet the needs of those already working in industry (either within waste management or in a complimentary discipline). In addition to defining the needs of industry in any curriculum, there are clearly other considerations that influence curriculum design, including academic standards and requirements; and the various limitations of industry and industry personnel (Davis, 2005). Indeed, the whole process of curriculum development is premised on complex interrelations among purposes, experiences, content and means of evaluation.

Whilst the author is not proposing a single curricula or class of membership for the waste management industry either nationally or internationally, there is little information available on the 'real' size of the sector and the full range of activities that fall within it across many countries, so it is not clear what constitutes the industry, its needs and skill requirements in many cases. There is a need to adequately quantify this sector for every economy, particularly if adequate education and training provisions are to be identified and the professionalism of the industry and the individuals working within it is to be recognised. If this does not occur, the waste management sector may risk losing core personnel to other sectors and limits opportunities for new recruitment.

7. References

Allen, J. (1963). Professional Identity. *Bios.* 34 (4): 167-170.

Bennett, R. (2000). The Logic of Membership of Sectoral Business Associations. Review of Social Economy. LVIII (1): 17-42.

CIWM. (2008). *"Still Not Taken the Plunge"*. CIWM: the Journal for Waste and Resource Management Professionals. The Chartered Institution of Wastes Management. November 2008. P17.

CIWM. (2009). *"Going Global"*. CIWM: the Journal for Waste and Resource Management Professionals. The Chartered Institution of Wastes Management. June 2009. P20-21.

Cochrane-Smith, M., Lytle, S. 1999 Relationships of Knowledge and Practice: teacher learning in communities. Review of Research in Education. Iran-Nejad, A & Pearson, C. (Eds)

Conn, W.D. (1993). Initiating the Development of an Integrated Waste Management Curriculum. *Resources, Conservation and Recycling.* 8. 55-61.

Davis, G. (2005). The Education and Training of Wastes Managers within the UK: Fulfilling the Educational Needs of Waste Managers Working in Small to Medium Sized Enterprises (SME's). ISWA 2005 World Congress, Buenos Aires, 7th-10th November 2005.

Davis, G. (2006). The Use of Case Studies for Integrating Sustainable Development into the Education of Engineers. *World Transactions on Engineering and Technology Education.* UNESCO International Centre for Engineering Education (UICEE), Monash University. Melbourne. 5 (1): 159-163.

Davis, G. (2008). Formulating an Effective Higher Education Curriculum for the Australian Waste Management Sector. *The Journal of Wastes Management.* 28: 1868-1875.

Davis, G (2009). Who Are Waste Managers: A Story of Constructed Identities and Professional Opportunities. Oral presentation at the 24th International Conference on Solid Waste Technology and Management, Philadelphia, US. 15-18th March 2009.

Davis, G., Read, A. (2007). Adding Value to the Waste Management Curriculum: Educating for New Technologies. Paper in proceedings of Sustainable Waste Management Conference. Melbourne. 19th-21st November 2007.

Day, C., Sachs, J. (2004). Professionalism, performativity, and empowerment: discourses in the politics, policies and purposes of continuing professional development. International Handbook on the Continuing Professional Development of Teachers (Eds Day, C and Sacks, J). McGraw Hill, Open University Press.

Doney, E. (1998). Developing Opinions: The Attitudes of ILS Staff to Continuing Professional Development. *Library Management*. 19 (8): 486-491.

Gallagher, K., Rose, E., McClelland, B., Reynolds, J., Tombs, S. (1997). People in Organisations: An Active Learning Approach. Oxford, Blackwell.

Hargreaves, D. (1994). The New Professionalism: The synthesis of professional and institutional development. *Teaching and Teacher Education*. 10(4). 423-38.

Harvey, R., Lemon, R., Southern, J. (1994). Personal Factors, Education and the Effectiveness of Professional Engineers. *European Journal of Engineering Education*. 19 (3): 351-363.

Heynes, P. (1994). CPD from the other side of the door. *Librarian Career Development*. 2(4): 13-15.

Hoyle, E. (1980). Professionalisation and de-professionalisation in education. World Yearbook of Education: The Professional Development of Teachers. (Eds Hoyle, E and IBISWorld, 2007. Waste Disposal Services in Australia Report. Industry Code Q9634. IBISWorld, January 2007.

LeanrDirect. (n.d). Annual Salary Guidelines for UK Public Sector Waste Personnel/Officers. Figures provided by Learn Direct, UK University for Industry, UK Government. www.learndirect-advice.co.uk. Retrieved 30th September 2009.

Megarry, J). Pp42-56. Kogan Page, London.

Inside Waste. (2009). "*Members have their say on WMAA*". Inside Waste. April 2009. P14.

Jackson, P. (1968). Life in Classrooms. Holt, Rinehart and Winston, New York.

London Economics. (2008). An Economic Impact Assessment of the CCPMO: Final report for the consultative committee for professional management organisations.

Maehr, D., Thompson, B., Mattfeld, G., Montei, K., Haufler, J., Kerns, J., Ramakka, J. (2002). Directions in Professionalism and Certification in the Wildlife Society. *Wildlife Society Bulletin*. 30 (4): 1245-1252.

Noon, P. (1994). CPD: Professional Development-Continuing and Compulsory? *Librarian Career Development*. 2 (4): 4-8.

Pullen, A., Beech, N., Sims, D. (2007). Exploring Identity: Concepts and Methods. New York, Palgrave Macmillan.

Royal Society of Chemistry. (2005). The Economic Benefits of Higher Education. Pricewaterhouse Coopers, January 2005.

Swank, W. (1987). Professionalism on the Wildlife Society. *Wildlife Society Bulletin*. 15: 55-67.

Tang, S., Choi, P.L. (2009). Teachers' Professional Lives and Continuing Professional development in Changing Times. *Educational Review*. 61 (1): 1-18.

UKCES. (2008) Empowering SSCs- Employer Driven Skills Reform Across the UK. A Re-Licencing Framework for the Sector Skills Councils. The UK Commission for Employment and Skills.

Walker, I., Zhu, I. (2001). The Returns to Education: Evidence from the Labour Force Surveys. Department for Education and Skills Research Report 313.

Wilson, K.M., Halpin, E. (2006). Convergence and Professional Identity in the Academic Library. *Journal of Librarianship and Information Science*. 38 (2): 79-91.

4

Institutional Development in the Urban Waste Market in Portugal. Market Structure, Regulation and Performance Analysis

Rui Cunha Marques and Pedro Simões
*Center of Urban and Regional Systems, IST, Technical University of Lisbon
Portugal*

1. Introduction

Many waste utilities managers and some think-tankers have considered the waste market as one of the best macroeconomic indicators of a country (US.EPA, 1999). They defend that the production of waste is deeply linked to the financial health of the citizens (O'Neil & Locke, 2004) and, therefore, to the financial health at a national level. Figure 1 shows the growth of urban waste in line with the Gross Domestic Product (GDP) and the population (OECD, 2002). The upward trend of waste production occurred at a rate slightly lower than GDP, but well above population growth.

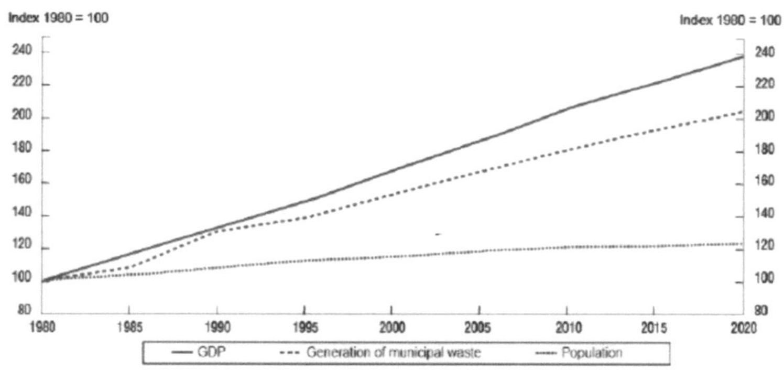

Fig. 1. Production of urban waste, GDP and population growth in OECD countries for the period 1980-2020

The increasing waste production brought several problems to Portugal as in the early 1990s, the country was not endowed with suitable facilities for waste treatment. The Portuguese Government at the time developed the first national strategic plan dedicated to urban waste treatment (acknowledged by PERSU I for the period 1997-2006), in order to fill this gap (Faria et al., 1996). PERSU I among other issues defined the Portuguese market structure for the final disposal of urban waste allowing for two types of utilities which cover the whole country, respectively the regional public companies and the regional municipal companies/systems. While the former comprises partnerships between the State and the

municipalities (public-public partnerships), the latter can establish partnerships with the private sector (of contractual type or institutionalized) or be managed by the association municipalities alone.
Notwithstanding the improvements achieved by PERSU I, other (stricter) needs came up together with society development. So, in 2006 the PERSU II was approved for the period between 2007 and 2016. It continues the policy of waste management taking into account the new (more ambitious) requirements formulated at national and Community level. In particular, it focuses on ensuring the compliance with the European Union (EU) objectives of deviating biodegradable waste from landfill and recycling and recovering packaging waste.
Only these points would represent more than enough reasons to study, analyze and understand this market. However, the waste sector is framed by many characteristics that increase the importance of doing research on it (Abduli & Nasrabadi, 2007). The provision of the urban waste service, as a public service, inherently associated with reduced incentives to be efficient and innovative; the continuous growth and development of societies in general, together with more and more waste produced that require the ultimate technological means; more robust management models (IRAR, 2008); and the relevance that this sector has gained in the municipalities budget are some of the features that justify this research.
In Portugal, the major responsibilities within this sector are assigned to the public sector, where the local administration has the ultimate responsibility for waste service (collection and treatment). Globally, the urban waste sector in Portugal is clearly divided into "primary", "secondary" and "tertiary" markets, encompassing, respectively, collection, treatment and recycling of urban waste (Massarutto, 2006).
The aspects previously referred to, along with the proliferation of private management in the urban waste sector and the growing concern with the protection of users' interest, led the Portuguese Government to create a regulator dedicated to this sector, the Institute for the Regulation of Water and Waste (IRAR). IRAR represented an innovative step in Europe, since only few countries have a regulator for the urban waste sector in a worldwide context. Besides the responsibility for the economic regulation of some segments of the waste sector, IRAR supervises the quality of service. It determines a set of 20 performance indicators for each operator and compares and displays publicly the results (sunshine regulation). Thus, the operators become "embarrassed" with a poor performance and try to correct the deviations. This approach has led to good outcomes by fostering the improvement of performance in the whole sector (Marques & Simões, 2008).
This article intends to present and analyse the urban waste sector in Portugal. In particular, we focus on the regulation of the Portuguese urban waste services and its results which is the most noteworthy feature of the Portuguese context. After this brief introduction, the current paper analyses the legal framework of the urban waste sector in Portugal. Next, section 3 discusses the major institutional framework of the Portuguese urban waste sector concerning the most relevant authorities and, in special, the sector-specific regulation. Section 4 shows the market structure of the urban waste services in Portugal, encompassing different aspects such as the ownership, the players, the integration of the sector, and the major figures that characterize the waste sector. The rules of the game, concerning the tariff setting, quality of service and public service obligations, are presented in section 5. Finally, some conclusions and ideas for future research are highlighted in last section.

2. Legal framework

The waste sector, in general, and the urban waste sector, in particular, has observed significant and positive changes in Portugal. The access and the quality of service provided improved significantly in few years. Not everything resulted in improvements for the sector, although we must highlight that the reforms introduced in the sector, mainly the ones after 1993, allowed for this recognised (even internationally) progress (Marinho et al., 2006).

These deep reforms started with the introduction of private capital participation in the provision of these services and with the vertical separation of the waste sector into the wholesale and retail services. However, the municipalities or the association of municipalities remained the ultimate responsible for the waste service (Decree-Law No. 239/97).

In Portugal, the urban waste sector is regulated by various legislative documents, which can be divided according to their scope, respectively into management and sector-specific legislation. The characterization of the management legal framework can start with Decree-Law No. 372/93. In this diploma, the legal regime of management of regional and municipal systems was established and the private capital participation was authorised in this sector (allowing for the participation of private companies and other similar entities, by means of concession contracts arrangements). This law also made a distinction between regional systems and single municipality systems. The first only encompassed, a priori, the waste treatment whereas the second comprised the collection. Recycling could be allocated to both systems.

The vertical separation of the sector was defined in detail by means of Decree-Law No. 379/93. Under this diploma, if the municipalities voluntarily accept it, the State directly operates and manages regional systems, or attributes these functions by means of concessions arrangements to public corporate entities or to companies resulting from the association of public and private entities, where the public sector must compulsorily hold the major stake, being the State responsible for the major investments.[1] Notice, however, that municipalities can gather themselves to manage the regional systems or to choose a private partner.

Decree-Law No. 294/94 established the legal regime of concession and management of regional systems for waste treatment. This diploma focused on the (juridical regime) concessions of waste regional systems introducing some important points (for example harmonizing the concession period, defining criteria for the tariff setting, etc).

An important legal landmark was the creation of an Observatory for the waste sector, through Decree-Law No. 147/95 which would be replaced by IRAR with the publication of Decree-Law No. 230/97. Decree-Law No. 362/98 established the statutes of IRAR, later expanded by Decree-Law No. 151/2002.

It is equally important to point out the regulation of the local corporate sector (SEL), and the possibility of implementing municipal companies, which occurred by means of Law No. 58/98 later replaced by Law No. 53-F/2006. The latter normative document states that SEL entities engaged in activities within the regulated sectors are subject to the authority of the

---

[1] This normative document was later on amended by Decree-Law No. 103/2003, which clarified the public interest assignments of the regional systems.

respective regulatory agency and that it is essential to hold public tenders to choose private partners when they exist.

Regarding the protection of users of essential public services, which includes the waste services, the Decree-Law No. 12/2008 had a great relevance. In terms of public service obligations, this diploma is a significant milestone for user protection. It aims to assure a minimum requirement set of principles, deemed to be indispensable for the quality of life in modern society.

Specifically for the waste sector, the legal framework starts with the publication of Decree-Law No. 239/97, where the rules that should be subject to waste management were established. Few years later, this diploma was altered by the Decree-Law No. 178/2006, which set the waste management regime, transposing into national law the EU Directives. In addition, after observing the operators' difficulties in disposing the material collected, the Portuguese government enacted this diploma that introduced the Organized Market for Waste (MOR), which is an economic tool, voluntary in nature, to facilitate and promote trade in the various waste streams, boosting their reuse or recovery by reintroducing in the economic cycle. It is intended that MOR centralizes in one place or system of the trading transactions of various waste streams, ensuring their rational allocation, reducing transaction costs and decreasing demand for primary raw materials.

Regarding the disposal waste treatment, Decree-Law No. 152/2002 established the legal regime that is subject to the issuing procedure for license, installation, operation, closure and post-closure maintenance of landfills for waste disposal. Order No. 209/2004 approved the European Waste List, which ensures the harmonization of existing legal requirements for the identification and classification of waste, thus making the economic agents aware of the legal system to which they are subject. Moreover, Decree-Law No. 1023/2006 defined the elements that must accompany the permit application of storage, sorting, treatment and disposal of waste.

From the need of establishing an emergency plan for urban waste in order to recover the delay in meeting the European targets for recycling and recovery of waste, both consistent with PERSU and European strategy, the Intervention Plan for Urban Solid Waste and Similar (PIRSUE) was approved. It is a tool of characterization and resolution of problems and mechanisms to guide the urban waste management. Later, the regulation on the integrated system for waste electronic registration was also approved.

More recently, Order No. 187/2007 came to approve the PERSU II - Strategic Plan for Solid Waste 2007-2016, which, as referred, represents the ongoing diploma focused on the urban waste management, in particular, the recycling targets, in accordance with European Directives.

Finally, the Decree-Law No. 127/2008 implement and ensure compliance within the obligations of the legal systems, regarding the creation of the European register of Pollutant Emissions and Transfer, and Hazardous Waste, and the prevention and control of pollution, in accordance with the European Directives.

3. Institutional framework

3.1 Responsibilities

The waste sector encompasses diverse kinds of entities that, in some way, have responsibility in the waste sector. In administrative terms, it is important to highlight the role of the regulatory agency (IRAR) and of other bodies of the Public Administration, namely, the Ministry for the Environment, Territorial Planning and Regional Development (MAOTDR), the Portuguese Environmental Agency (APA), the Directorate-General for Consumer Affairs (DGC), the Competition Authority (AC), and the municipalities. Apart from being involved in waste service provision, the public company *Empresa Geral de Fomento* (EGF), a sub-holding of *Águas de Portugal*, plays a very relevant and structural role in the environment domain. Finally, there is the *Sociedade Ponto Verde* (SPV) which performs activity in the tertiary market.

IRAR is the sector-specific regulator for the urban waste services (and water sector as well). Within the scope of its main functions, IRAR supervises and regulates the waste utilities in charge of the urban waste collection and treatment corresponding to concession arrangements. Its objectives include safeguarding the quality of service provided and supervising and ensuring the equilibrium and sustainability of the sector. IRAR suggests to the MAOTDR the tariffs that should be adopted by the regulated utilities. In fact, the MAOTDR represents the highest entity responsible for the diverse policies of the environmental sector, including the waste sector.

In Portugal, the environment agency (APA) gained more importance due to the merging between the Institute for the Environment and the Institute for Waste. APA, established in 2007, has the objective of creating conditions for a greater effectiveness in the management of environmental policies and sustainable development.

The DGC is a public institute created to promote the policy of user rights, as well as to coordinate and implement measures aimed at user protection, information and education and to support user organisations. In cooperation with IRAR, this body plays a significant role in protecting users from the eventual abuse of the waste services.

The AC, which was created in 2003, has transversal powers over the Portuguese economy to apply rules for competition, in cooperation with the sector-specific regulatory bodies. The AC's mission is to ensure that the rules of competition are implemented in Portugal, according to the principles of a market economy and free competition, with a view to ensuring that markets work efficiently. It seeks to achieve a high level of technical progress and secure greater benefits for users. The AC collaborates with IRAR in all matters related to competition in the waste market.

The municipalities are responsible for providing the waste services. In this context, they can manage themselves the urban waste services or delegate them to private entities by means of concession arrangements or to other entities, such as parishes and user associations. In the case of treatment services, municipalities and their associations might carry out themselves this task (or delegate it to a private company) or accept a partnership with the state to constitute a regional company (by means of a concession agreement) and in this way take advantage of the investment being mostly made by the State (EGF).

The EGF, which is a State owned company, is the main corporate group in the waste sector in Portugal. Its mission is to contribute towards resolving national problems in the waste service domain (treatment and collection), within a framework of economic, financial, technical, social and environmental sustainability. Presently, it encompasses 13 wholesale waste companies in the scope of its activities.

SPV is a private, non-profit making organization that was set up in November 1996 associated with the Green Dot program, to promote the selective collection, sorting, take-back and recycling of packaging waste.

3.2 Regulation

In Portugal, there is a sector-specific regulator (IRAR) which has responsibility for waste systems (and municipal and regional water and wastewater systems) that have been delegated as concession arrangements. Among its main objectives are the protection of user rights, the guarantee of sustainability and of economic viability of the waste operators.

IRAR is an authority governed by public law, endowed by administrative and financial autonomy, with legal personality and with its own patrimony but subject to the MAOTDR supervision. The board of directors is also politically appointed (for small mandates of 3 years). However, it is financially independent since it is entirely funded by the regulated concessionaires by means of a (regulatory) tax.

At the beginning, the objective was to consolidate the regulatory model, since it started in 2003 (Baptista et al., 2003). This model is now fully operational and its strategy is based on two broad planes of intervention: one at the level of the structural regulation of the waste sector, being concerned with the restrictions to the entry of operators and with the level of service integration (Marques & Simões, 2008) and the second at the level of regulating the behaviour of operators, known as conduct regulation.

In the conduct regulation scope, IRAR's strategy includes regulating the actions of the operators concerning economic aspects and the quality of service provided. Economic regulation is the most important way of regulating the behaviour that is allowed for operators, insofar as monopolistic prices tend to be higher than the prices practised in competitive markets. In this regard, the regulatory model is somewhat limited. As far as the regional systems are concerned, IRAR issues recommendations on the tariffs proposed by the concessionaire every year but they are always subject to the approval of the MAOTDR. Concerning the municipal concessions (basically on water sector), IRAR only issues opinions in the public tender documents and on the design draft of the contracts. It does not intervene in the process of setting tariffs, which is a decisive element while choosing concessionaires, except in situations of economic and financial rebalancing. The tariffs are regulated by the terms of the contract signed.

In terms of quality of service regulation, IRAR opted for discussing and publicising the results of the operators' performances. This model is known as sunshine regulation. The objective of this regulatory approach is to "embarrass" the operators that perform poorly, so that they will more likely correct their weaknesses (name and shame policy). Although this method does not set tariffs and its coercive power is limited, the display and public discussion of the behaviour of the regulated operators has very positive effects, inducing competition among operators and leading to progressive performance improvements in the whole market (Marques, 2005).

This model lies in the publication of an annual report with performance scores, obtained on the basis of a set of performance indicators applied to the operators. The preparation of this document, which is based on benchmarking, includes a joint assessment of performance, with comparisons between the operators, and an individual assessment of the performance of each operator in qualitative and quantitative terms (Marques, 2006). In addition to its attributions, IRAR also provides some recommendations/observations to the operators on the results obtained per each indicator. In this scope, the awareness that IRAR seeks to achieve, concerning the activities of the operators, is developed by means of pressure by users and citizens in general, through their protection groups, the media, the political classes (government/ political parties) and NGOs.

Recently, the Decree-Law No. 277/2009 replaces the IRAR by the Regulatory Authority for the Water and Waste Services (ERSAR). Besides the new name, which seeks to clarify that its action is specifically about the water and waste services and not generally on waste and on water as a resource, the new legislation reinforces the sector's regulation. This results in the extension of the activity of ERSAR to all water and waste utilities of these services (also the ones that do not correspond to the concession arrangements), as well as wider responsibilities and powers.

4. Market structure

4.1 Services ownership and management

The waste sector in Portugal has been historically associated with the public management. However, after the legal change (due to Decree-Law No. 379/93, allowing private participation), the sector started to observe a proliferation of the private sector (Pinela et al., 2003), mainly in the last decade, not only at a concession level, but also concerning the urban waste services provision, for example, the refuse collection and the urban cleaning (mainly through short-term contracts, between one and five years). Table 1 presents all possible management arrangements of the waste sector, although some of them not exist in the Portuguese reality (IRAR, 2008) yet.

	Management models of State ownership	
	Utility	Partnership
Direct Management	State	No one
Delegated	Public company	No one
Concession	Regional public company	Public-Public[2]
	Management models of municipal ownership	
	Utility	Partnership
Direct Management	Municipality	No one
	Semi-autonomous utilities	No one
	Municipal associations	Public-Public (several municipalities)
Delegated	Municipal (or regional) companies	No one or Public-Private
	Local corporate entities	No one or Public-Private
	Parishes and users associations	Public-Public (several municipalities)
Concession	Municipal companies	Public-Private (municipalities and private companies)

Table 1. Management models in urban waste sector

[2] This kind of partnerships is established between State and municipalities, with the possibility of evolving to Public-Private Partnerships (State, municipalities and private companies).

4.2 Players in the sector

4.2.1 Wholesale segment

As stated, the waste sector in Portugal is structured in two segments, namely the wholesale[3] and the retail segment. The wholesale companies have usually a regional scope. At this stage, there are 29 regional systems in Continental Portugal, which are split into regional public companies, and regional municipal utilities which in turn are divided into municipal associations (AM), municipal concessions (celebrated between AM and private companies) and regional companies (which include only the municipalities as partners).

Entity	Business Model	Control/Concessionaire
ALGAR	Concession	EGF, S.A.
AMARSUL	Concession	EGF, S.A.
BRAVAL	Concession	AGERE, EM, from Braga
ERSUC	Concession	EGF, S.A.
REBAT	Concession	EGF, S.A.
RESAT	Concession	EGF, S.A.
RESIDOURO	Concession	EGF, S.A.
RESIESTRELA	Concession	EGF, S.A.
RESIOESTE	Concession	EGF, S.A.
RESULIMA	Concession	EGF, S.A.
SULDOURO	Concession	EGF, S.A.
VALNOR	Concession	EGF, S.A.
VALORLIS	Concession	EGF, S.A.
VALORMINHO	Concession	EGF, S.A.
VALORSUL	Concession	EGF, S.A.
AMBILITAL	Regional municipal company	AMAGRA (AM)
AMBISOUSA	Regional municipal company	VALSOUSA (AM)
ECOBEIRÃO	Regional municipal company	AM Planalto Beirão
ECOLEZÍRIA	Regional municipal company	RESIURB (AM)
GESAMB	Regional municipal company	AMDE (AM)
RESIALENTEJO	Regional municipal company	AMALGA (AM)
TRATOLIXO	Regional municipal company	AMTRES (AM)
AMAVE	Association of Municipalities	
AMCAL	Association of Municipalities	
AMVDN	Association of Municipalities	
LIPOR	Association of Municipalities	
RESITEJO	Association of Municipalities	
ZAGOPE	Private Company	A. Gutierrez / AM Raia-Pinhal
FOCSA	Private Company	FCC / AMDSFE, AMTQT and AMTNFT via RDN

Table 2. Waste utilities operating in the Portuguese wholesale segment

[3] Wholesale segment is related to the service of waste treatment, providing the last destination to the waste, whether it is disposal or other types (corresponding to the secondary and tertiary markets mentioned before).

Regional public companies include 15 concessions arrangements which had been set between the central government and the respective companies, except one, Braval, whose majority shareholder is the municipal company of the city of Braga (AGERE). The other ones are owned (at least 51%) and controlled by EGF. As to the AM's, there are currently 12 in charge of urban waste treatment, whereas only 5 of them operate on their own behalf. The remainders operate through regional municipal companies. The last management model is the municipal concession, which is a simple long-term contract celebrated by a group of municipalities (AM in this case) with a private company. There are only two cases of these in Portugal. Table 2 shows the Portuguese waste management in the wholesale segment.
Regarding the tertiary market (recycling and reselling), the major and oldest entity is SPV, which supports the selective collection and sorting and is also responsible for the take-back of waste likely to be recycled. Nowadays, there are also other operators responsible for other streams (Pássaro, 2003), such as Amb3E for electronic equipment, Valorpneu for used tires and Ecopilhas for used batteries.

4.2.2 Retail segment

The part of the system which deals directly with the user is called retail segment. Its main activity is the refuse collection but it can also encompass the selective collection. Other services of its scope are relative to particular activities of collection of large volumes or even with urban cleaning.
The retail service coincides usually with what is called the primary market, where the service directly provided by the municipality remains the most representative. These account for about 76% of the total served population. However, most of them contract-out their services awarding short-term contracts (1 to 5 years). Other part of the population is served by semi-autonomous utilities, which are endowed with financial and administrative autonomy and control 5% of the market. Municipal companies (covering 12% of the population) can be split into two types: the ones fully owned by the municipality (66%) and the ones with mixed capital (34%). In both types the municipality is the major shareholder and they are always corporate companies. There are also some particular cases where the wholesale utility is responsible for the waste collection. The market structure of Portuguese waste collection is presented in table 3.

Arrangement	Urban waste collection	
	Number	Population
Directly by municipalities	218	7 510 528 (76,2%)
Semi-autonomous utilities	5	490 674 (5,0%)
Municipal companies	15	1 199 321 (12.7%)
Regional utilities	6 (40 municipalities)	596 371 (6,1%)
TOTAL	**244 (278 municipalities)**	**9 851 424**

Table 3. Urban waste market structure in Continental Portugal for the retail segment

4.3 Integration of the sector and other services

In Continental Portugal, the waste services, with few exceptions, generally are not vertically integrated, i.e. the wholesale and retail systems are provided by different operators. In horizontal terms, the operators have a limited degree of integration and (even after excluding the parishes) there are 267 operators for almost 10 million inhabitants.

In the wholesale service the rule is to provide solely the waste service, but there are exceptions. Regarding the retail waste services, it is more common to see operators providing also other services, such as water and wastewater services and transportation.

4.4 Numbers of the sector

Whereas in the PERSU I, for 2005, the aim for landfill would be 25% of the urban waste produced, the result in 2007 was 63% (APA, 2008). At the same time, this value represents one of highest percentage of landfilling in EU-15 (Defra, 2007), as it is presented in figure 2. This was mainly related to the lack of equipment available and how it is distributed across the country. There are only two incinerators in Portugal, one in Lisbon and the other in Oporto. The same happens for organic recovery equipments which are also in reduced number at this stage.

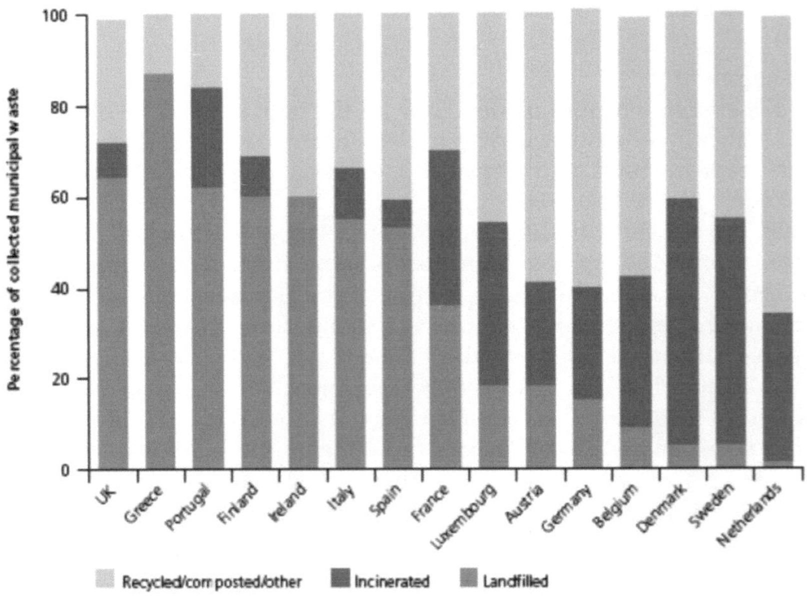

Fig. 2. Performance management of municipal waste in EU-15 (2005)

In 2006, in Portugal 4 641 103 tons of urban waste were collected. Of this, 89,5% corresponds to the refuse collection and only 10,5% fell within the collection of selective multi-material and biodegradable waste. In the following year there was an increase of about 4 698 774 tons of urban waste collected, which corresponds to 1,27 kg per inhabitant and day, a value below the EU average. Despite this, there was a decrease in selective multi-material collection of about 2%. Figures 3 and 4 present the evolution of waste production and selective collection in Portugal, respectively (APA, 2008).

Between 1995 and 2006, and compared both with the EU-15 and the current EU-27 Member States, the annual urban waste production per capita in Portugal has always remained below the European average production per capita.

Concerning the physical characterization of the urban waste, presented in figure 5, the biodegradable fraction corresponds to more than a quarter of the traditional waste composition, 36%. This figure highlights the need to give priority to organic recovery,

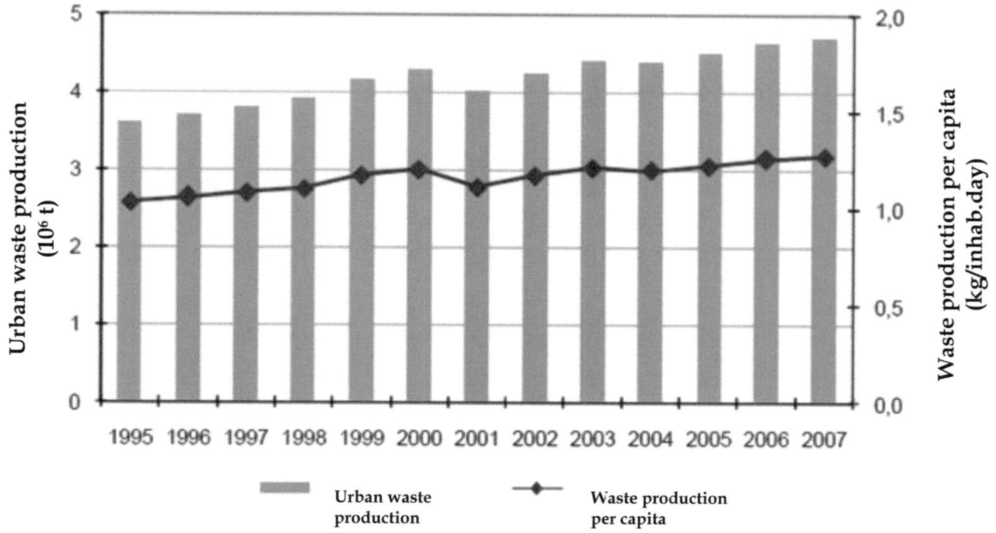

Fig. 3. Total and per capita waste production between 1995 and 2007

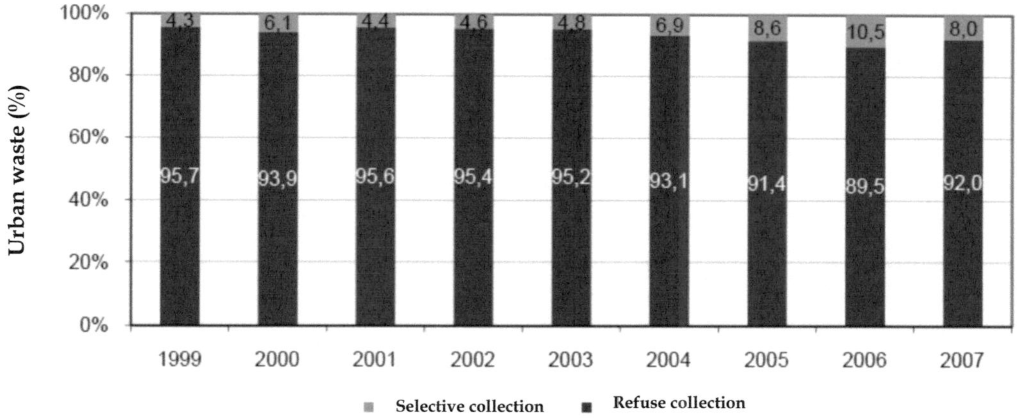

Fig. 4. Refuse and selective collection between 1995 and 2007

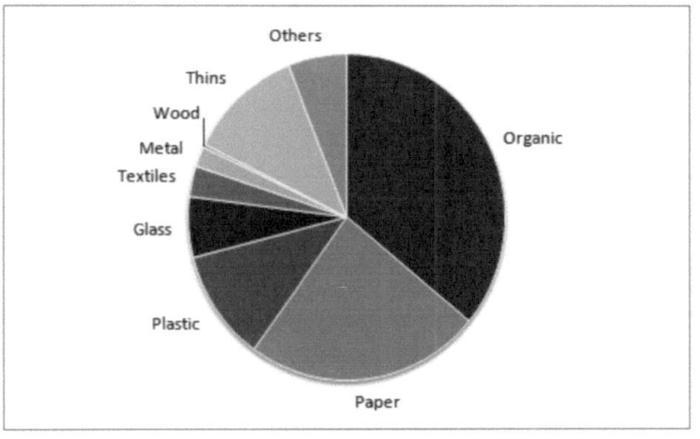

Fig. 5. Typical characterisation of the urban waste (in terms of percentage)

recycling, and incineration with energy recovery, against the current trend of landfill (INETI, 2009).
In terms of waste infrastructures, the Portuguese reality (until December 2007) is indicated in Table 4 (APA, 2008).
At this time, the national average of people per drop-off container was 322. In the region of Lisbon and Tagus Valley, this average was 364 inhabitants per drop-off container.
Figure 6 displays the main destinations of the urban waste in Portugal. The landfill is still the most frequent. With the implementation of PERSU I, the waste dumps were completed in 2001.

Infrastructures		Predicted	In construction	Working	Total
Landfill		2	2	34	38
Organic valorisation		11	2	8	21
Incineration		0	0	2	2
Transfer stations		0	2	76	78
Selective collection	Sorting facilities	1	2	26	29
	Drop-off centre	7	0	185	192
	Drop-off container	40	-	28 723	28 763

Table 4. Portuguese infrastructures for waste sector

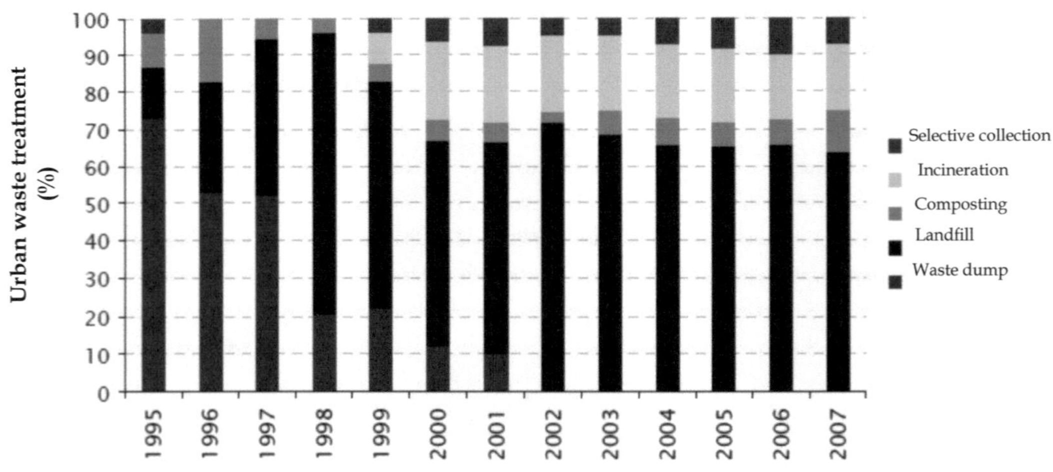

Fig. 6. Types of urban waste treatment in Portugal

APA (2008) identified a positive relation between GDP and waste production in Portugal. This relation provides evidence of producing waste with the wealth generated by translating a form of eco-efficiency at the national level. It reveals that the intensity of urban waste produced per unit of wealth in Portugal grows slightly in the period under review. In Portugal 35,88 kilograms of waste were produced per 1000 € of GDP in 2006 (APA, 2008).

Figure 7 also allow us to observe the relative amplitude (APA, 2008) between waste production and household expenses and compare their patterns over time. Nationally, between 1995 and 2007, both the urban waste production and the GDP increased (about 29% during 1995-2006), highlighting some connection between them.

5. Rules of the game

5.1 Tariff setting

In Portugal, the establishment of prices and tariffs is related to the model of waste management in question. IRAR does not have any kind of functions regarding municipal services (directly provided by municipality), semi-autonomous services and the different models of municipal companies. Thus, it is the responsibility of the local administration, in this case the Municipal Assembly, to approve the tariff systems proposed by the Municipal Executive or by the board of directors of the semi-autonomous services. Regarding the regional services, the assembly is responsible for approving tariffs. In the case of municipal (or regional) companies, it is possible to find different situations, according to the nature of the company in question. If it is an institutionalized PPP, then the General Assembly approves the tariffs, proposed by the Board of Directors. Concerning other municipal companies, the Town Hall or the Board of Directors approves tariffs.

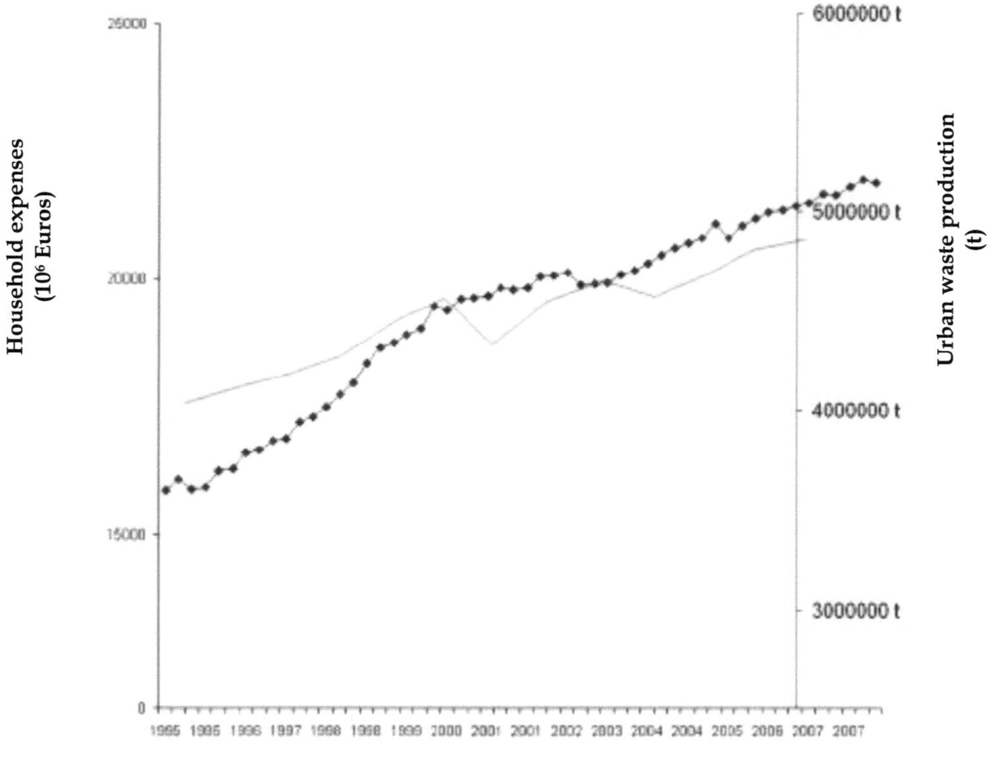

Fig. 7. Evolution of household expenses and the urban waste production

In municipal systems involving concession arrangements, tariffs are defined in the concession contract signed between the concession granting authority (Town Hall) and the concessionaire. The setting of tariffs is established in the winner bid (public tender), under the terms of the Decree-Law No. 147/95. IRAR cannot directly interfere in tariff setting, unless the economic and financial equilibrium of the concession is jeopardised by unpredictable reasons when the contract is signed. In such situation, IRAR can be invited to issue its opinion on the matter.

In regional systems, IRAR has other kind of powers, based on an investment programme that has been defined previously as well as on the annual account report of the companies and the respective budget. The formula for defining the tariff system proposed in the concession contracts of regional systems consists of a hybrid methodology, based on the rate of return regulation method established contractually and introducing a mechanism to share gains in productivity. Although it varies from contract to contract, the rate of return is fixed by the concession granting authority and consists of a base rate (Treasury Bonds or Euribor) plus a risk premium of 3%, applied to the capital stock achieved and the legal reserve.

Regarding the retail segment, it must be noted that each waste service has their own classification of users and their own tariff blocks (associated with the water service), which sometimes vary a great deal between operators (in some cases a free service). A study carried out by CESUR (2004), based on a survey to the municipalities, the waste charges vary with a) the supply (or not) of tap water, b) type of consumer (domestic, industrial, etc.), c) water consumption, d) percentage of water bill, e) type of collection system, f) frequency of collection, and g) municipality characteristics. In addition, it diverges in their implementation, that is, through a fixed part, a variable part, or both in the same tariff system. Beyond this disparity of tariff systems, the lack of sustainability of them is another concern of IRAR. Figure 8 presents this reform.

In this regard, IRAR, in 2009, issued a recommendation (since it is not entirely compulsory) intending to harmonise not only the structure of the tariff system but also its values, and to introduce some principles concerning the social tariffs. It is common to find social tariffs and special tariff systems for large families.

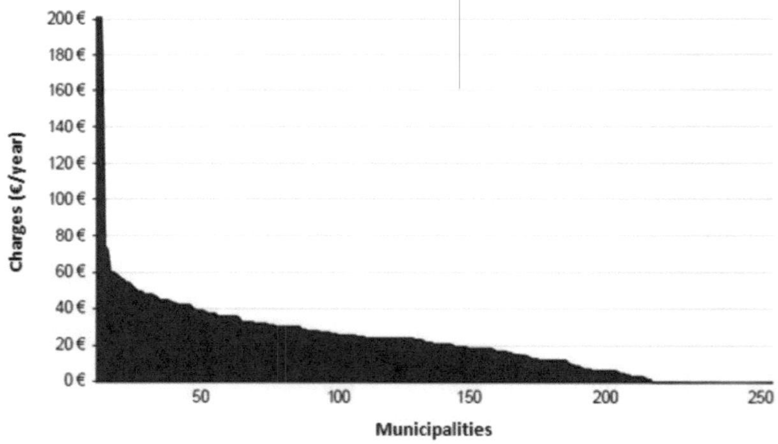

Fig. 8. Distribution of annual average tariff paid for refuse collection

5.2 Quality of service

The quality of service regulation became one of most important attributions of IRAR. In this scope, the quality of waste (and also water) service is regularly monitored. Actually a very high level of service has been observed since then. This positive evolution is a consequence of the role of the sector-specific regulator, which encourages the quality of service amongst the waste utilities regulated. This atypical situation, even in the worldwide context, has recognised the waste sector in Portugal as an example to follow by other countries (e. g. Italy and Brazil).

IRAR carries out an annual benchmarking exercise, in which it establishes, compares and publicises the performance of the regulated operators. For this purpose, the regulator has

PERFORMANCE INDICATORS	2004	2005	2006	2007
Protection of the user interests				
User service accessibility				
a) Service coverage (%)	100	100	100	100
b) Selective collection coverage (%)	67	77	79	80
c) Average waste charges (€/ton)	26.6	24.6	26.6	26.7
Quality of service supplied to users				
d) Answers to written complaints (%)	41	89	92	97
Sustainability of the operator				
Operator's economical and financial sustainability				
e) Operating cost coverage ratio (-)	1.5	1.58	1.65	1.6
f) Unit running costs (€/ton)	21.6	22.6	25.36	25.64
g) Solvency ratio (-)	0.19	0.51	0.55	0.59
Operator's infrastructure sustainability				
Recycling (%)	4.1	6.0	6.4	7.3
i) Organic recovery (%)	2.8	2.9	1.5	2.5
j) Incineration (%)	66.1	82	79	68
k) Waste landfill (%)	74.5	89.6	88	78
l) Landfill utilisation (%)	128	116	121	122
Operator's operational sustainability				
m) Failure in heavy duty equipment (nr./10^3 ton/year)	0.17	0.16	0.14	0.15
n) Waste characterisation (-)	1.5	2.8	2.3	1.7
Operator's human resource sustainability				
o) Employees (nr./10^3 ton/year)	0.35	0.46	0.49	0.52
Environmental sustainability				
p) Leachate tests performed (%)	90	80	96	96
q) Leachate quality upon treatment (%)	86	79	87	89
r) Utilisation efficiency of energy resources (kWh/ton)	-66.5	-90.4	-87.9	-73.9
s) Monitoring of groundwater quality (%)	84	97	90	84
t) Monitoring of air quality (%)	99.8	99.5	100	100

Table 5. Performance indicators for solid waste services used by IRAR and their results

defined a set of 20 performance indicators, focused on the wholesale segment of this service. These groups have also been divided into three categories, namely indicators aimed at protecting user interests, the sustainability of the operator and the environmental sustainability. Table 5 summarises the performance indicators defined by IRAR and shows the average results for the period 2004-2007 (four years). The results are generally very positive, although there are some indicators that have worsened. Nevertheless, this had to do with changes in the definitions of these indicators and a greater rigour in implementing them.
In addition, IRAR issues comments concerning the results of each indicator for every regulated operator. The performance obtained is compared with reference values (optimum values, or close to them, which IRAR deems to be reasonable and attainable by the waste utilities), considering the factors of the operational context in which each utility acts. From this assessment, IRAR classifies the performance of each operator in qualitative terms, taking into account the quality of service provided, whether it is poor, average or good, through a red, yellow or green ball per indicator, respectively, as it is presented in figure 9.
The annual regulatory process carried out by the IRAR results in the publication of an annual report (Portuguese Water and Waste Sector Report - RASARP) where the benchmarking results are presented and discussed. This document is a key element of the regulatory model adopted. Indeed, RASARP plays an important role in promoting a greater effectiveness and efficiency of the operators and provides reliable and easy to interpret information to users and other stakeholders about the waste services.
Together with the reformulation of its statutes and regulatory power (replaced by ERSAR), IRAR is developing a new and updated set of performance indicators in order to foster incentives to provide a service with better quality and attain other goals. For instance, in the Portuguese waste sector, the indicator coverage started not making much sense, since 100% was achieved by the waste utilities.

5.3 Public service obligations

Public service obligations are a fundamental tool for the protection of user interests. In Portugal they are considered in the "Waste Law" (Decree-Law No. 178/2006, which altered the Decree-Law No. 239/97) and the Essential Public Services Act (Decree-Law No. 12/2008), in which basic principles, such as, the universality or the continuity of the waste services are, respectively, guaranteed.
The MOR introduced some important principles, such as the equal access to market, costs, transparency and accuracy of the information provided and security for transactions, as well as the standards for the protection of the environment and public health. The creation of a MOR should encourage the participation of investors and of the waste producers themselves.
The Essential Public Services Act outlines various aspects to uphold the user interests: the right to participation (organisations representing users have the right to be consulted while defining regulations for the legal framework governing public services), the obligation to provide information when public services supply is suspended (except for unexpected situations or *force majeure* reasons, the service cannot be suspended without suitable warning) and the right to partial quittance (the payment of a public service cannot be

Benchmarking

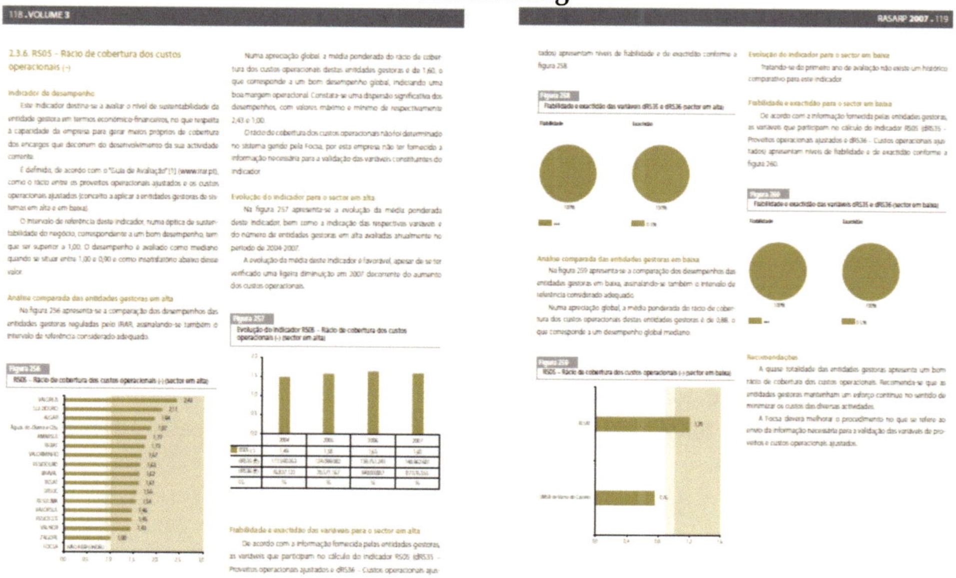

Sunshine regulation

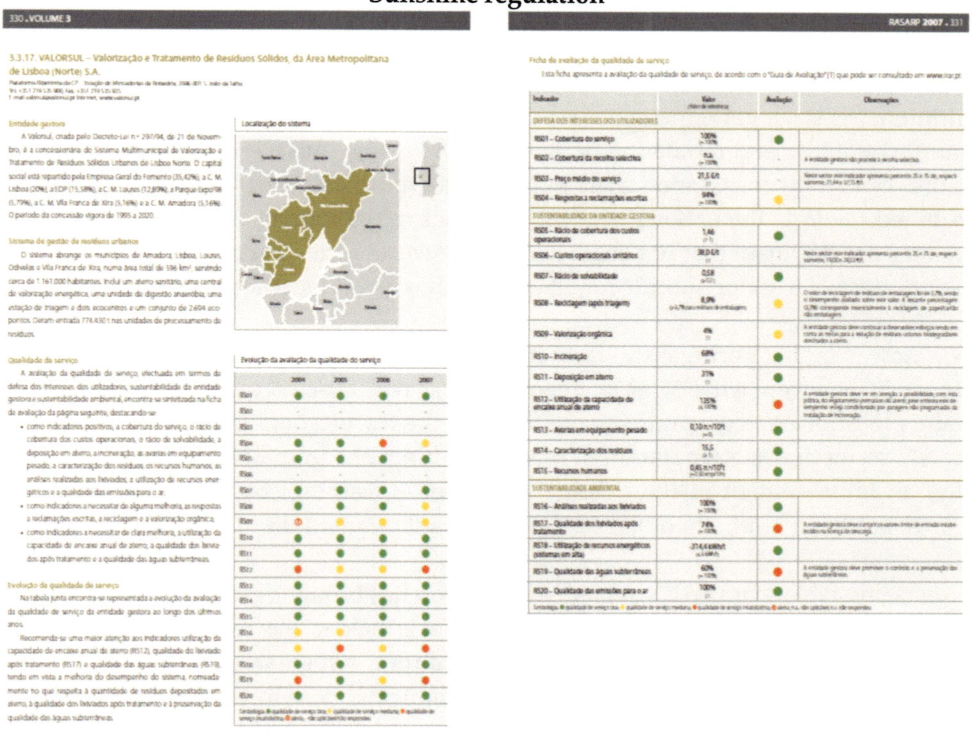

Fig. 9. Benchmarking and sunshine regulation of IRAR

refused). Furthermore, it comprises the assurance of quality standards (any service provided must conform to high quality standards), minimum consumption (imposing and charging for minimum consumption is forbidden), billing (users are entitled to receive a bill specifying the sums presented), prescription by lapse of time (the right to require payment lapses after a period of six months after the service has been provided), the injunctive nature of rights (any agreements or dispositions that exclude or limit the rights of users by this law are considered to be null and void) and safeguards entitlements (all legal dispositions that specifically prove to be more favourable to users are safeguarded).

With the aim of improving transparency and raising awareness about costs in the waste sector, the Local Finance Act (Decree-Law No. 2/2007) established that both the tariff systems and a complete breakdown of the costs of the service should be made available on the website of the municipality.

In addition, operators are compelled to have a complaint book and send all complaints to IRAR within a period of ten working days in order to guarantee a greater efficiency while handling user complaints.

6. Conclusions

In Portugal, in the past two decades the waste sector has evolved considerably. Taking into account the situation of this sector in the early 1990s and the current panorama, it is easy to understand the great contribution of the legal framework and of the large investments that were allocated to the waste sector. Moreover, the establishment of dedicated regulation represented an important (additional) role to this development. The significant increase of refuse collection coverage, beyond the percentage of suitable treatment, is one of the relevant signs of this progress.

In addition, the large participation of stakeholders by promoting recommendations and the technical credibility and reputation of the operators along with their freedom from political pressure, especially considering their institutional vulnerability, are also factors that have contributed towards the national and international recognition of the vast operational and regulatory development that has characterised the waste sector in Portugal, particularly due to the presence of a sector-specific regulator, which is an atypical situation in the worldwide waste sector.

Notice, however, that Portugal still faces important challenges in this area, namely, the great discrepancy between what it is charged and the real cost of the service, the need for operators to improve their productive efficiency, the clarification of the role of the State in the waste sector and the separation of regulatory functions from operational functions and, finally, an improvement of the existing legislation governing this sector which, in some cases, is outdated and, in other instances, is still limited in the context of the new requirements.

It is unquestionable that IRAR's sunshine regulation has brought great improvements to the waste sector, despite its little coercive power. Nevertheless, the recent reform both of its statutes and regulatory model, turning it into a stronger regulator, is seen as an optimal window to develop even more the waste sector, inducing more incentives to waste utilities, focusing on a service with better quality, simultaneously sustainable and affordable for the users.

7. References

Abduli, M. & Nasrabadi, T. (2007). Municipal solid waste management in Kurdistan Province, Iran. *Journal of Environmental Health*, Vol. 69, No. 7, pp. 51-55

APA (2008). *Relatório de Estado do Ambiente 2007*. Portuguese Association for the Environment (APA), Lisbon, Portugal

Baptista, J.; Pássaro, D. & Santos, R. (2003). *As Linhas Estratégicas do Modelo de Regulação a Implementar pelo IRAR*. Institute for the Regulation of Water and Waste, Lisbon, Portugal

CESUR (2004). *Os sistemas Tarifários de Resíduos Sólidos Urbanos em Portugal*. Center of Urban and Regional Systems (CESUR), Instituto Superior Técnico, Technical University of Lisbon, Portugal

Defra (2007). *Waste Strategy for England*. Presented in LGA Waste Strategy Debate Event, London, United Kingdom

Faria, A.; Chinita, A.; Ferreira, F.; Presumido, M.; Inácio, M. & Gama, P. (1996). *Plano Estratégico para os Resíduos Sólidos Urbanos (PERSU)*. Ministry for the Environment, Lisbon, Portugal

INETI (2009). *Prevenção de Residuos Urbanos – Proposta de Programa*. National Institute for Engineering, Technology and Innovation (INETI), Lisbon, Portugal

IRAR (2008). *Relatório Anual do Sector de Águas e Resíduos em Portugal*. Institute for the Regulation of Water and Waste (IRAR), Portugal

Marinho, A.; Didelet, F. & Semiao, V. (2006). Municipal solid waste disposal in Portugal. *Waste Management*, Vol. 26, No. 12, pp. 1477-1489

Marques, R. & Simões, P. (2008). Does the sunshine regulation approach work? Governance and regulation model of the urban waste services in Portugal. *Resources Conservation and Recycling*, Vol. 52, No. 8-9, pp. 1040-1049

Marques, R. (2005). *Regulação de Serviços Públicos*. Sílabo Press, Portugal

Marques, R. (2006). A yardstick competition model for Portuguese water and sewerage services regulation. *Utilities Policy*, Vol. 14, No. 3, pp. 175-184

Massarutto, A. (2006). Waste management as a service of general economic interest: is the self-sufficiency principle still justified? *Working Paper Series in Economics*, Universitá di Udine and Iefe, Udine, Italy

O'Neil, R. & Locke, E. (2004). Institutional and financial consideration in regional solid waste management. *Water, Air, & Soil Pollution*, Vol. 4, No. 2, pp. 227-236

OECD (2002). *Towards Sustainable Household Consumption? Trends and Policies in OECD Countries*. Organisation for Economic Co-operation and Development (OECD), Paris, France

Pássaro, D. (2003). Report: waste management in Portugal 1996 and 2002. *Waste Management*, Vol. 23, No. 1, pp. 97-99

Pinela, A.; Madeira, L.; Levy, J. & Teles, M. (2003). *Mercado dos Resíduos em Portugal*. Associação das Empresas Portuguesas para o Sector do Ambiente (AEPSA), Lisbon, Portugal

US.EPA (1999). *National Source Reduction Characterization Report for Municipal Solid Waste in United States*. United States Environment Protection Agency, Washington D. C., United States of America

5

Composting Barrel for Sustainable Organic Waste Management in Bangladesh

Md. Azizul Moqsud
Institute of Lowland Technology, Saga University
Japan

1. Introduction

Inadequate collection and uncontrolled disposal of solid waste cause a serious health hazard to inhabitants and environment (Goyal et al., 2005). So, solid waste management has become a major environmental problem confronting urban areas in developing countries (Pfammtter & Schertenleib, 1996). Sujauddin et al. 2008 shows that household solid waste can be converted from burden to resource in developing countries. Like other developing countries, the major portion (about 84%) of the total solid waste is organic in Bangladesh (Moqsud et al., 2005a). In composting process, the organic portion of wastes is converted into stable end product of compost by bio-degradation (Hong, 2005; Rahman, 1993). As a result, the volume, weight and moisture content are reduced; the potential odor and the pathogens minimized, and nutrients are available for agricultural production (Witter, 1998; Shinha, 2001). The spread of diseases is minimized due to the destruction of some pathogens and parasites at elevated temperatures (Golueke, 1972, Vesiland & Rimer, 1981, Bhide & Sundarsan, 1983).

Like any tropical and subtropical regions, Bangladesh has got a large amount of rainfalls and considerable sunlight throughout the year. But it is important to maintain limited amount of moisture content in the compost process for effective composting operations.

At moisture content > 65 %, anaerobic condition and generation of bad odors as well as the depletion of the aerobic bacteria will be induced (Goyal et al., 2005). At moisture content < 40%, bacterial activity for decomposition is reduced (Rahman, 2004). Hence, it is necessary to design a special type of composting reactor to cope with the heavy rainfall in these regions for composting. The in-vessel composting reactor requires small area for installation and operation which is essential in the densely populated urban areas (Iyengar & Bhave, 2005). The problems of fly breeding, odor generation, and rodents are eliminated because of the enclosed aerobic design of the reactor (Chang, et al., 2006). The conventional steel barrel, (Figure 1 a), has not become popular as a composting reactor in Bangladesh as it can not skip over the bad odor generation and also takes a longer time to complete the composting process (Moqsud, 2004). So a modification is needed to overcome these drawbacks of the conventional composting barrel. This chapter presents an assessment of a modified steel barrel used as a composting reactor in Bangladesh, which is suitable, eco-friendly and effective in the typical sub-tropical monsoon climate. Field tests were performed in order to examine the effectiveness of the proposed in-vessel composting reactor. Benefit-cost ratio of this modified barrel composting plant is analyzed to ensure the financial viability.

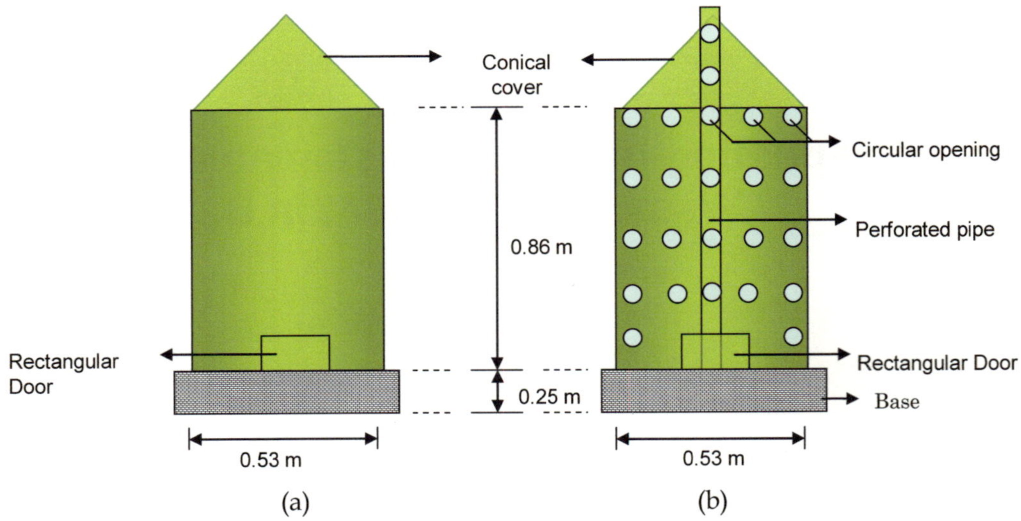

Fig. 1. Schematic of a composting barrel: (a) conventional type (b) modified type

2. Materials and methods

Figure 1 (b) shows the conventional as well as the proposed modified barrel used in this study. The modified barrel made of steel has a circular shape with a diameter of 0.53 m and height of 0.86 m. This barrel, generally available in the market to store oil and chemicals, was collected from the recycle shops and modified. The modification include:

- 0.0125 m diameter openings throughout the surroundings to ensure aerobic digestion inside the barrel

The conventional composting barrel has no opening in the surroundings and it has no perforated PVC pipe in the middle portion of the barrel which caused almost anaerobic condition inside the barrel, and consequently a bad odor generation during the composting operation (Moqsud, 2004). A conical-shaped cover made of steel is set at the top to avoid infiltration of rain water and to protect from excess heat and sunlight into the organic waste both in the conventional and the newly modified composting barrel. A 0.25 x 0.30 m rectangular opening with a hinge system was made at the bottom side for collecting the composting mass from the barrel both in the conventional and the proposed modified barrel (Figure 1). The both barrels were placed on a raised base to manage leachate resulting from the composting process (Moqsud, 2003). The outside of the barrel was painted green for aesthetic and corrosion protection. The average cost of each conventional and modified barrel was around US$25 and US$29, respectively. As the modification was done by using the recyclable PVC pipe and created the openings by using the hand drill machine the cost for modification is not so prominent. The organic waste was separated and collected from the source of generation in the study area. This biodegradable waste comprised of mainly kitchen garbage of raw vegetables and some cooked food waste. There were some biodegradable paper waste as well as fruits leftovers which were homogenized by cutting to approximately 0.05-0.08 m in length (Iyenger, Bhave, 2005; Moqsud, 2003). The large size organic wastes are pulverized into small pieces for quick composting process (Chang et al., 2006; Hong, 2005). Then these homogenized organic solid wastes were deposited in the composting barrel by removing the conical cover shown in Figure 1.

The basic physico-chemical properties of the organic wastes of the study area are summarized in Table 1.

Bulk density of the collected waste (kg/m^3)	300
Mean daily waste generation rate (kg/person/day)	0.27-0.38
Moisture content (%)	56 %
pH	7.2
C/N ratio	21

Table 1. The basic properties of the collected organic wastes in the study area

The change in waste volume, temperature, pH and moisture content were monitored during composting. After 45 days, the decomposing organic wastes were collected from the bottom rectangular opening of the barrel and layered in the sunlight for maturing for one week.

One of the main objectives of composting is to reduce the volume of the organic waste. Waste reduces volume from its initial value during the composting process. The total volume of the waste was calculated everyday by measuring the height of the waste and was compared with the previous day's volume.

Temperature is an important controlling parameter for proper composting operation (Sundberg, 2004). Variation in temperature was measured daily during composting by inserting thermometer (range 0^0 to 100^0 C) 0.1 m above from the bottom through the circular side-openings in the barrel. The reading was equilibrated for 5 minutes.

Measurements of pH in the compost were done by adding 0.010 kg of compost sample (taken from 0.10 m above the bottom of the barrel) in 0.0001 m^3 distilled water and mix thoroughly for several minutes (Sundberg, 2004; Jackson, 1973). The pH was measured by a digital pH meter (JENWAY -3051) by inserting the probe into the water-compost mix solution.

The variation of moisture content was determined by gravimetric procedure of weighing samples before and after the water is removed. The sample was weighted first to get its wet weight. Then it was dried for 24 hours in a 110°C oven. After weighting the sample and subtracted the weight of the container the dry weight was achieved. The difference between the wet and dry weights was the weight of the water that the sample originally contained. By dividing this water weight by the wet weight gave the moisture content as a fraction. The following equation (1) was used for determining the moisture content of the compost.

$$\text{Moisture content (\%)} = \{(\text{wet weight-dry weight})/\ \text{wet weight}\} \times 100 \qquad (1)$$

The nutrient content of the finished compost sample was determined by standard methods of chemical analysis (Goyal, 2005; Sundberg, 2004; Jackson, 1973). Nitrogen content was determined as total Nitrogen by Kjeldhal's method. Phosphorus content was measured as total Phosphate (acid digest) using vanadomolybdophosphoric acid method, and Potassium (acid digest) from various compost samples were determined using flame photometry (Systronics) fitted with element specific filters.

3. Results and discussion

The variation of volume change of organic solid waste is shown in Figure 2. It is seen that after composting was completed in about 35-40 days, the volume of the organic waste

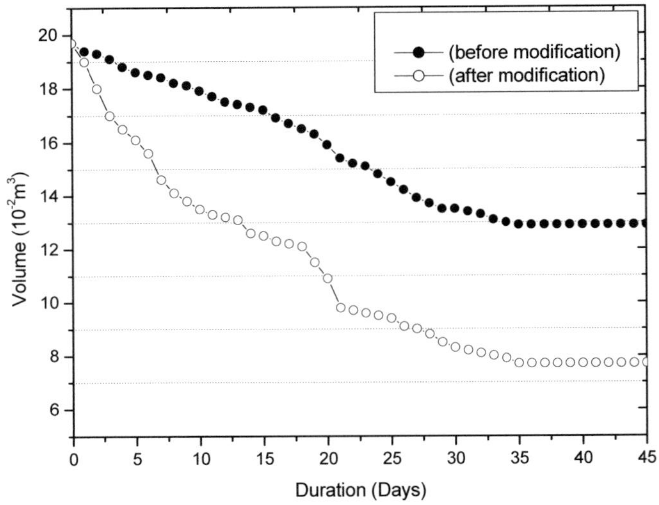

Fig. 2. Variation of waste volume during composting before and after the modification of the barrel

became 50-70 % of the original volume. The volume reduction was quicker, when the modified composting barrel was used.

After 4 weeks, the volume became 70% and 50% less of its original volume for the case of after and before modification, respectively. This is so as the aerobic condition was maintained in the middle as well as the surroundings of the new modified barrel. Microorganisms reproduced very rapidly at early stage of composting due to the abundance of easily-degradable organic matter. For this, the rate of volume reduction is usually fast (Zheng, 2004).

Figure 2 shows that during the initial days (3-7 days), the rate of volume reduction was quick, while after 14 days it became slow and almost constant after 21 days and 35 days in the modified and conventional barrel, respectively. This observation is mainly due to the quantity of nutrients for the bacteria were depleted, and thereby the rate of decomposition became slower in the later stage. In the case of conventional barrel, the volume of the organic waste reduces slowly. This is due to the lack of aerobic condition inside the barrel and the presence of excess amount of moisture content (Figure 4).

The temperature variation during composting is illustrated in Figure 3. It shows that temperature rises from the initial day to the maximum at 21 days and then it begins to fall and became constant after 37 days.

The broad range of optimum temperatures for composting process is from 45 to 65^{0}C (Sundberg, 2004).This allows a large variety of micro-organisms to participate in the process. However, if the peak temperature is more than 72°C, some useful bacteria, which are responsible for the aerobic decomposition, will die (Ahmad, 2000).

The temperatures generated by the mesophilic and thermophilic bacterial activity lead to the destruction of diseases causing microorganisms. During the initial days, the increased temperature in the composting barrel was mainly caused by the more exothermic reactions associated with the respiratory metabolism of the microorganisms involved in the composting operation (Sundberg, 2004; Tchobanglous, 1977). But at that time, the moisture content is high, which prevents to reach at the peak temperature. Again, it is found that after the modified composting barrel generated a peak temperature of 69°C in contrast to 60°C before modification.

In case of the conventional barrel, moisture content is generally high as there were not much available openings to remove or evaporate the moisture of the leachate and thereby lowering the temperature inside the reactor. The durations of the optimum temperature for composting are about 22 days and 28 days for the case of before modification and after modification of the barrel, respectively. The longer length for the modified one would be favorite to the composting process, even though it has a drawbacks in terms of temperature >65°C.

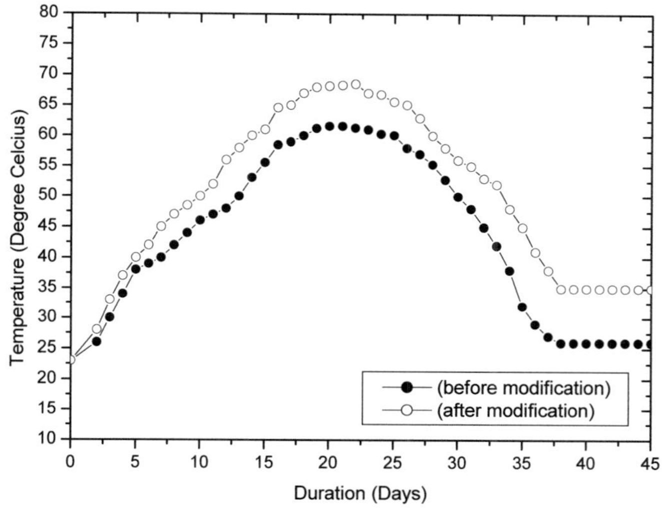

Fig. 3. Variation of temperature during composting before and after modification of the barrel

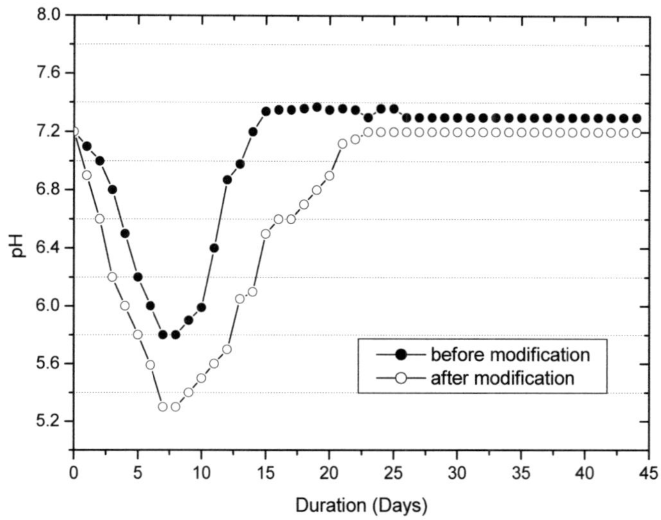

Fig. 4. Variation of pH during the composting before and after the modification of the barrel

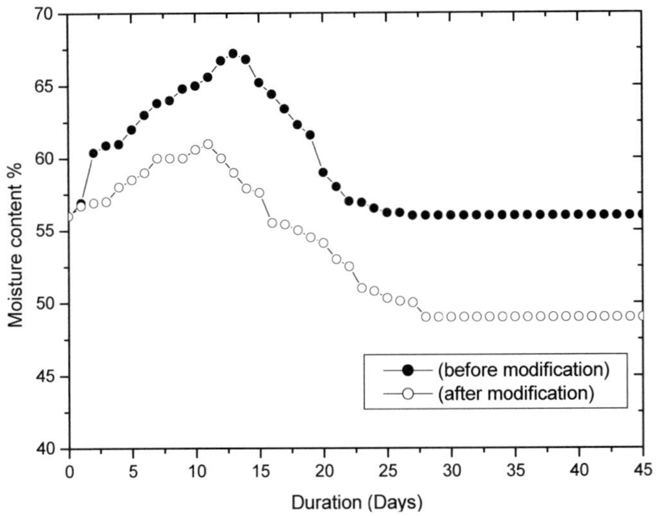

Fig. 5. Variation of moisture content during composting before and after modification of the barrel

The variation of pH with time during the composting process is plotted in Figure 4. This variation directly indicates the extent of decomposition within the compost mass. The optimum pH range for most bacteria is between 6.0 and 7.5 (Parks et al., 1988). During the initial period (first 2 to 7 days), pH decreased to 6.0 or less during the initial phase of temperature rise.

The duration of pH > 6.0 for the case of modified barrel is longer than the conventional barrel. Therefore, it may imply that modified barrel has higher performance. From day 12, pH value of the compost began to rise to about 7.4. Initially, the microorganisms reproduced very rapidly because there was large portion of easily-degradable organic matter in the wastes. With the activity of the microorganisms, considerable amount of organic acids (acetic acid, butyric acid etc.) were produced, and consequently, the pH value of the compost decreased (Li & Zhang, 2000). With volatization of the organic acids, the pH value of the compost increased as more decomposition of organic matter had occurred (Fang & Wong, 1999).

Moisture content was determined by weight loss of compost samples, which were oven dried at 110 °C for 24 h (Dresboll & Kristensen, 2005). At moisture levels above 68 %, water begins to fill the interstices between the particles of the wastes, reducing the interstitial oxygen and causing anaerobic conditions with a resultant offensive odor (Rahman, 1993; Tachobanglous, 1977). When the moisture contents drop much below 40%, the composting process becomes slow (Goyal et al, 2005). Figure 5 shows the variation of moisture content with duration of composting.The maximum moisture content reached nearly 67% before modification and about 60% after the modification of the barrel. So, by ensuring the aerobic condition in the middle and the surroundings of the barrel, the peak moisture content can be maintained at the optimum level for proper composting. As in the initial stage (5-12 days), the bacterial activities and the decomposition of organic waste were more significant. The values of moisture content were also high in that period and it decreased after some days (15-25 days) as the bacterial activities decreased at that time. However, the moisture content

in the conventional barrel is higher than the desired moisture content due to lack of aeration in all stages of the composting operation than that of the modified barrel.
The result of chemical analysis of compost sample is presented in Figure 6.

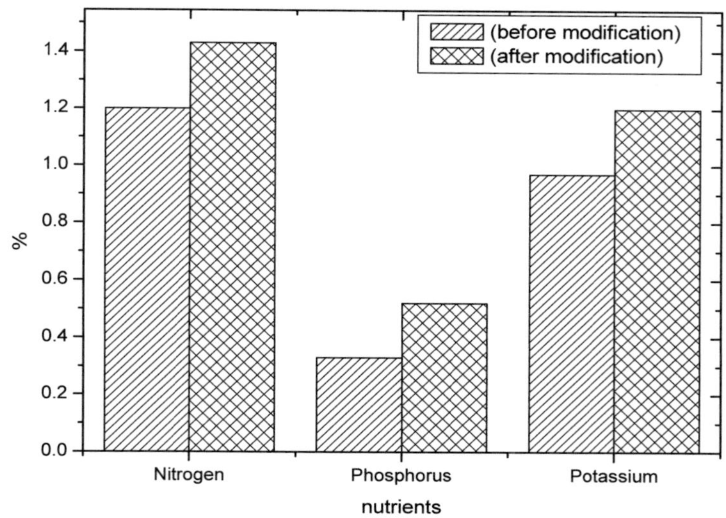

Fig. 6. Quantity of different nutrients present in the compost sample

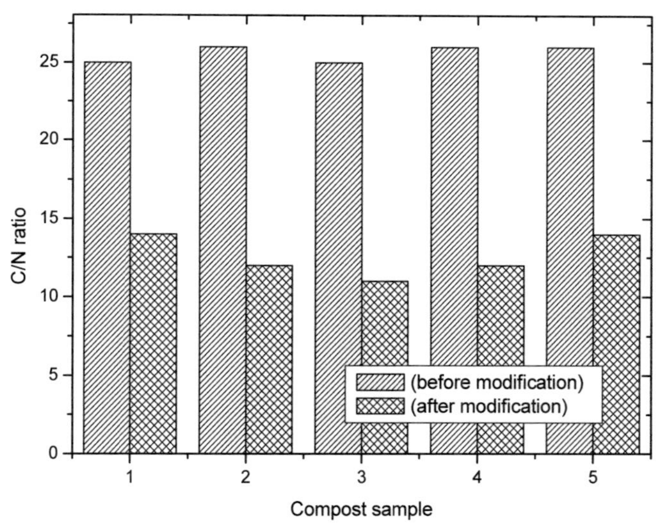

Fig. 7. Carbon- Nitrogen ratio of ready compost samples

As the ultimate goal of the composting of organic solid waste is to use the compost as a soil conditioner and also as a fertilizer in the agricultural field, it is important to examine the values of different nutrients. All chemical analyses were performed according to the standard methods of soil and compost analysis (Goyal, 2005; Sundberg, 2004; Jackson, 1973). It is observed that the values of nutrients i.e. Nitrogen, Phosphorus and Potassium (NPK) were very much similar as reported in other countries (Asija et al., 1984). The NPK values were lower than the ideal values (N=1.5%, P=1.2%, K=0.8%) when the conventional barrel was used because of the lack of aeration during the composting (Verma et al., 1999).

Decomposition of organic matter is brought about by microorganisms that use the carbon as a source of energy and nitrogen for building cell structure. More carbon than nitrogen is needed. If the excess of carbon is too great, decomposition decreases when the nitrogen is used up and some of the organisms die (Nakasaki et al., 2005, Polprasert, 1996). The stored nitrogen is then used by other organisms to form new cell material. Figure 7 shows that the carbon-nitrogen (C/N) ratio of the ready compost varies from 11 to 14 in different samples in the study area after the modification. In the case of conventional barrel reactor the C/N ratio was found to be higher (above 24) than the recommended values (12-16). The compost from the conventional barrel would not be suitable for agricultural land application since the excess carbon would tend to utilize nitrogen in the soil to build cell protoplasm, consequently resulting in loss of nitrogen in the soil on which it would be applied.

4. Financial assessment of modified barrel composting project

The generation of solid waste was found to increase almost linearly with increasing of per capita income. Figure 8 shows the variation of the waste generation rate with the variation of per capita (person) income of selected low and middle-income family in the study area. When the per capita income per month is US$6-8, per capita waste generation is about 0.27 kg/day and when per capita income per month is US$67-75, per capita waste generation is about 0.38 kg/day.

Three different revenues were assessed from the modified composting barrel plant. These are

- fees charged by the collection scheme to the service beneficiaries (households) on a monthly basis (approximately US$0.3).
- revenues from the sale of compost (US$0.08 /kg) and revenues from the sale of recyclable materials like hard plastics, card board, glass and metals.

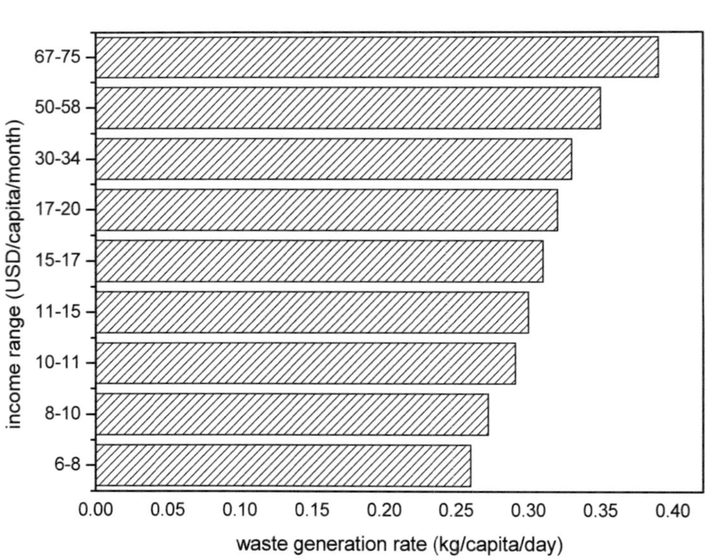

Fig. 8. Variation of per capita waste generation rate with respect to per capita income

Item *Cost*	*US$/year*	*Item* *Revenue*	*US$/year*
Depreciation cost for collection Investment (life time 5 years)	2422	Collection fees	3000
		Compost sales	12458
Operation cost for collection and composting	10152	Recyclables	333
Total	12574	Total	15788

Table 2. Yearly costs and revenues of modified composting barrel plant (including collection)

Table 2 gives the summary of costs and revenues for a modified composting plant of capacity of 1.865 tons/day on a yearly basis. It is seen that the plant is financially viable when operating at 1.865 tons/day. It is evident that the revenues from the collection fees are partly cross-subsidizing the composting activities. Hence, it seems advisable to combine composting activities with neighborhood waste collection to ensure a viable operation of the scheme. An additional advantage of a combination of waste collection and composting is the direct influence on improving waste composition for composting in the collection area, as continuous contact with the customers is available and appropriate information may be disseminated (e.g. promoting source separation and separate collection). The depreciation was calculated using a lifetime of 5 years and interest rate of 15%. The cost items comprise barrel modification plant set up, salaries and uniforms of the employees both for collection and in the composting plant, maintenance of collection vehicles, and expenses for electricity and water. Total revenues from the sale of the recyclables such as hard plastics, cardboards, glass and metal are US$333. The benefit-cost ratio of the modified composting barrel plant is > 1. Financial analysis confirms the results of other investigations on decentralized urban composting plants, showing that small-scale plants struggle with their economic viability if all costs have to be covered by the plant revenues (Lardinois & Furedy, 1999). However, our results show that a plant of capacity 1.865 ton/day may be viable in the study area where the rent for land is relatively smaller than the capital city as land acquisition in urban areas is always one financial key obstacle for initiating a composting plant. The decentralized waste collection and composting activity relieves a certain burden from municipal budgets in the study area (Zurbrugg et al., 2005). The municipal waste transportation and landfill costs can be reduced approximately by US$9500 per year. This estimate takes into account that the composting plant reduces the amount of waste, which needs to be transported by municipal trucks as well the reduction of the municipal expenses for its final disposal. With or without municipal support, any composting plant should however focus on long-term financial feasibility where operational costs are covered by revenues. Therefore, marketing strategies and the development of a market for compost are crucial for the long term success of a composting plant (Zurbrugg et al., 2005).

5. Conclusions

Reduction of waste volume was faster in the modified composting barrel than the conventional barrel reactor. The volume becomes 50% and 70% of its original volume before and after modification of the composting barrel, respectively after 4 weeks. The barrel composting was operated in the mesophilic and thermophilic temperature bend, which was very effective for proper composting operation. The quality of compost in terms of C/N ratio is better in the modified composting barrel than the conventional barrel. Nutrient concentrations of compost, produced in the modified composting barrel, were also satisfactory. The biochemical quality of the compost produced in the modified composting barrel was found suitable. The benefit-cost ratio for large scale modified composting barrel plants is more than 1. Thus, the modified composting barrel can be an eco-friendly, efficient and a sustainable solution of organic waste management alternative in Bangladesh.

6. References

Ahmed, M. F., Rahman, M. M., 2000. Water Supply and Sanitation: Rural and low income urban communities, ITN-Bangladesh, Center for Water Supply and Waste Management, BUET, Dhaka, Bangladesh.

ASija, A. K., Pareek, R. P., Singhania, R. A., Singh, S., 1984. Effect of method of preparation and enrichment on the quality of manure. Journal of Indian Society of Soil Sci. 32, 323-329.

Bhide, A. D., Sundersan B. B., 1983. Solid waste management in developing countries, Indian National Scientific Documentation Center, New Delhi.

Chang, I. J.; Tsai, J. J.; Wu, H. K., 2006. Thermophilic composting of food waste. Bioresource Technol. 97, 116-122.

Dresboll, B. D., Kristensen, K. T., 2005. Delayed nutrient application affects mineralization rate during composting of plant residues. Bioresource Technol. 96, 1093-1101.

Fang, M., Wong, J. W. C., 1999. Effect of lime amendment on availability of heavy metals and maturation in sewage sludge composting. Journal of Environmental Pollution. 106(1), 83-89.

Golueke, C. G., 1972. Biological Reclamation of Solid Wastes. Rodale Press, Emmanus, USA.

Gantzer C., P. Gaspard, L. Galvez, A. Huyard, N. Dumouthier, J. Schwartzbrod. 2001. Monitoring of bacterial and parasitological contamination during various treatment of sludge. Wat. Res. 35(16), 3763-3770.

Goyal, S., Dhull, S. K. and Kapoor, K. K., 2005. Chemical and biological changes during composting of different organic waste and assessment of compost maturity. Bioresource Technology, 96, 1584-1591.

Hong, J. H., Park, K. J., 2005. Compost biofiltration of ammonia gas from bin composting. Bioresource Technology, 96, 741-745.

Iyengar, R. S., Bhave, P.P., 2006. In-Vessel composting of household wastes. Waste Management. 26, 1070–1080.

Jackson, M. L., 1973. Chemical analysis of soil. McGraw Hill Publications Company.

Li, G. X., Zhang, F. S., 2000. Solid waste composting and production of fertilizer. Chinese Chemical Industry Press, Beijing, P R China.

Moqsud, M. A., 2003. A study on composting of solid waste. A Thesis of Masters of Science in Environmental Engineering, Bangladesh University of Engineering and Technology, Dhaka, Bangladesh.

Moqsud, M. A; Rahman, M. H., 2004. Composting of Kitchen garbage in Bangladesh. Proceedings of the sixth international summer symposium, Japan Society of Civil Engineering, Saitama, Japan, p. 413-416.

Moqsud, M.A; Rahman, M.H., 2005. Biochemical quality of compost from kitchen garbage of Bangladesh. Proceedings of the 20th International conference on Solid Waste Technology and Management, Philadelphia, p. 440-447, USA.

Moqsud, M.A, Rahman, M.H, Hayashi, S., Du, Y.J., 2005a. An assessment of modified composting barrel for sustainable waste management in tropical regions. 4th International conference on Environmental Informatics, July 26-28, Xiamen, China. p. 130-136.

Nakasaki, K.; Yaguchi, H., Sasaki, Y., Kubota, H., 1993. Effects of pH control on composting of garbage. Waste management and Research. 11(2), 117-125.

Pfammatter R., Schertenleib R., 1996. Non-governmental refuse collection in low-income urban areas. Lessons learned from selected in Asia, Africa and Latin America. SANDEC Report No.1/96.Water and sanitation in developing countries EAWAG/SANDEC, Duebendrof , Switzerland.

Polprasert, C., 1996. Organic waste recycling-technology and management. Wiley, Chichester, west Sussex, England.

Rahman, M. H., 1993. Recycling of solid waste in Bangladesh. The International Journal of Environmental Education & Information.UK, 12(4), 337-342.

Rahman, M. H., 2004. Composting of Solid waste in Bangladesh. Proceedings of the 19th international conference on Solid Waste Technology and Management, Philadelphia, USA p 45-49.

Sinha, A. H. M. M. and Enayetullah, I., 2001. Solid waste management with Resource recovery options. Proceeding of the International Conference on Professional Development Program 4, Center for Environmental and resource Management, Dhaka, 2-4 February, p. 32-37.

Sujauddin, M., Huda S. M. S, Hoque, R., 2008. Household solid waste characteristics and management in Chittagong, Bangladesh. Waste Management 28, 1688-1695.

Sundberg, C., Smars, S., 2004. Low pH as an inhibiting factor in the transition from mesophilic to thermophilic phase in composting. Bioresource Technol. 96, 746-752.

Tchobanoglous, G., 1977. Solid Waste: Engineering Principles and Management Issue, McGraw Hill Publications Company, New York.

Verma, L. N., Rawat, A. K., Rathore, G. S., 1999. Composting process as influenced by the method of aeration. Journal of Indian Society of soil sci. 47 (2), 368-371.

Vesilind, P. A., Rimer, A. E., 1981. Unit operations in resource recovery engineering, Prentice-Hall, Inc, New Jercy.

Witter, E., Lopeaz-Real, J. M., 1988. Nitrogen losses during the composting of sewage sludge and the effectiveness of clay soil, zeolite and compost in adsorbing the volatilized ammonia. Biological Wastes 23, 279-294.

Zheng, G. D., Chen, T. B., 2004. Dynamic of Lead specialization in sewage sludge composting. Journal of Water Sci. and Technol. 50(9), 75-82.

Zurbrugg, C., Drescher, S., Rytz, I., Sinha, A. M., Enayettullah, I., 2005. Decentralized composting in Bangladesh, a win-win situation for all stakeholders. Resources conservation and recycling. 43 , 281-292

6

Solid Waste Management through the Application of Thermal Methods

Konstantinos Moustakas and Maria Loizidou
National Technical University of Athens,
School of Chemical Engineering,
Unit of Environmental Science & Technology
9, Heroon Polytechniou Street, Zographou Campus, Athens
Greece

1. Introduction

Human life in modern societies is inevitably related to waste generation. Around 255 million tones of municipal solid waste were generated in the 27 Member-States of the European Union in 2006, an increase of 13% in comparison to 1995. This represented an average of 517 kg of municipal waste per capita, an increase of 9% over 1995. Therefore, it is not strange that waste management has become a crucial subject with increasing interest for scientists, local authorities, companies and simple citizens.

The effective management of solid waste involves the application of various treatment methods, technologies and practices. All applied technologies and systems must ensure the protection of the public health and the environment. Apart from sanitary landfill, mechanical recycling and common recycling routes for different target materials, the technologies that are applied for the management of domestic solid waste include biological treatment (composting, anaerobic digestion) and thermal treatment technologies (incineration, pyrolysis, gasification, plasma technology).

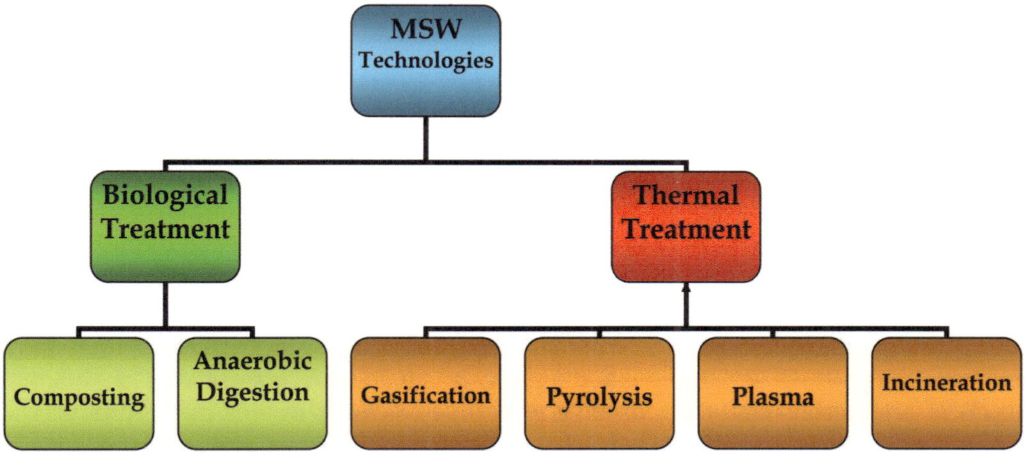

Fig. 1. Different biological and thermal methods for solid waste management

This chapter focuses on the description of the alternative thermal practices for municipal solid waste management. Thermal methods for waste management aim at the reduction of the waste volume, the conversion of waste into harmless materials and the utilization of the energy that is hidden within waste as heat, steam, electrical energy or combustible material. They include all processes converting the waste content into gas, liquid and solid products with simultaneous or consequent release of thermal energy.

According to the New Waste Framework Directive 2008/98/EC, the waste treatment methods are categorized as "Disposal" or "Recovery" and the thermal management practices that are accompanied by significant energy recovery are included in the "Recovery" category. In addition, the pyramid of the priorities in the waste management sector shows that energy recovery is more desired option in relation to the final disposal.

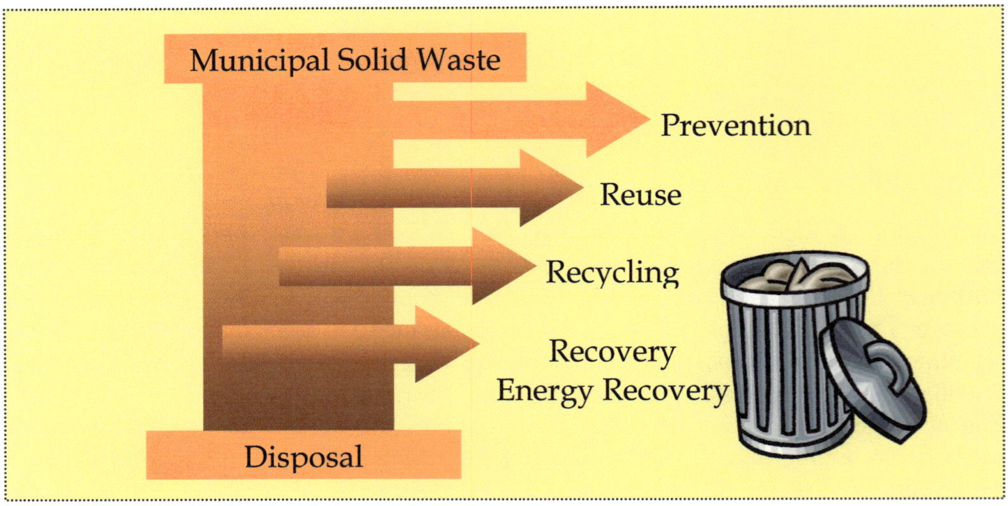

Fig. 2. Pyramid of the priorities in the waste management sector

That is why more and more countries around the world develop and apply Waste-to-Energy technologies in order to handle the constantly increasing generated municipal waste. Technologically advanced countries in the domain of waste management are characterized by increased recycling rates and, at the same time, operation of a high number of Waste-to-Energy facilities (around 420 in the 27 European Member-States). More specifically, on the basis of Eurostat data the percentages of municipal waste treated with thermal methods for the year 2007 in Denmark, Sweden, Luxembourg, Netherlands, France (Autret et al., 2007), Germany, Belgium and Austria were 53%, 47%, 47%, 38%, 36%, 35%, 34% and 28% respectively. On the other hand, there are still Member-States that do not apply thermal techniques in order to handle the generated municipal waste, especially in the southern Europe and the Baltic Sea. Such countries include Bulgaria, Estonia, Iceland, Cyprus, Latvia, Lithuania, Slovenia, Malta, Poland, Romania and Greece.

General information about the use of thermal technologies for solid waste management around Europe and worldwide is provided. Data referring to incineration - mass burn combustion, pyrolysis, gasification and plasma technology is presented. The different aspects of each technology, the indicative respective reactions, as well as the products of each thermal process, are described. The issue of air emissions and solid residues is addressed, while the requirements for cleaning systems are also discussed for each case.

Finally, the first attempt to treat municipal waste in Greece with the use of gasification / vitrification process is presented.

2. Incineration

2.1 General

The incineration (combustion) of carbon-based materials in an oxygen-rich environment (greater than stoichiometric), typically at temperatures higher than 850°, produces a waste gas composed primarily of carbon dioxide (CO_2) and water (H_2O). Other air emissions are nitrogen oxides, sulphur dioxide, etc. The inorganic content of the waste is converted to ash. This is the most common and well-proven thermal process using a wide variety of fuels.

During the full combustion there is oxygen in excess and, consequently, the stoichiometric coefficient of oxygen in the combustion reaction is higher than the value "1". In theory, if the coefficient is equal to "1", no carbon monoxide (CO) is produced and the average gas temperature is 1,200°C. The reactions that are then taking place are:

$$C + O_2 \rightarrow CO_2 + 393.77J \quad (1)$$

$$CxHy + (x + y/4)\,O_2 \rightarrow xCO_2 + y/2\,H_2O \quad (2)$$

In the case of lack of oxygen, the reactions are characterized as incomplete combustion ones, where the produced CO_2 reacts with C that has not been consumed yet and is converted to CO at higher temperatures.

$$C + CO_2 \;\; +172.58J \rightarrow 2CO \; (3)$$

The object of this thermal treatment method is the reduction of the volume of the treated waste with simultaneous utilization of the contained energy. The recovered energy could be used for:

- heating
- steam production
- electric energy production

The typical amount of net energy that can be produced per ton of domestic waste is about 0.7 MWh of electricity and 2 MWh of district heating. Thus, incinerating about 600 tones of waste per day, about 17 MW of electrical power and 1,200 MWh district heating could be produced each day.

The method could be applied for the treatment of mixed solid waste as well as for the treatment of pre-selected waste. It can reduce the volume of the municipal solid waste by 90% and its weight by 75%. The incineration technology is viable for the thermal treatment of high quantities of solid waste (more than 100,000 tones per year).

A number of preconditions have to be satisfied so that the complete combustion of the treated solid waste takes place:

- adequate fuel material and oxidation means at the combustion heart
- achievable ignition temperature
- suitable mixture proportion
- continuous removal of the gases that are produced during combustion
- continuous removal of the combustion residues
- maintenance of suitable temperature within the furnace
- turbulent flow of gases
- adequate residence time of waste at the combustion area (Gidarakos, 2006).

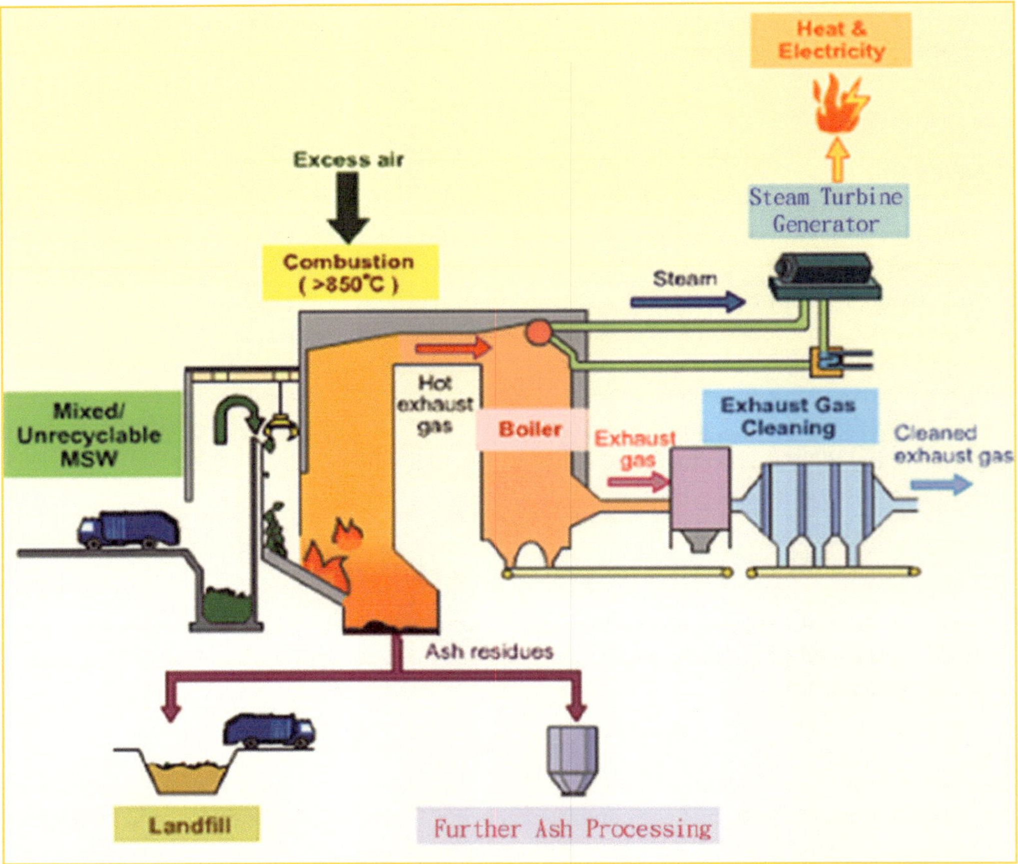

Fig. 3. A schematic diagram of incineration process

The existing European legislative framework via the Directive 2000/76/EC prevents and limits as far as practicable negative effects on the environment, in particular pollution by emissions into air, soil, surface water and groundwater, and the resulting risks to human health, from the incineration and co-incineration of waste (European Commission, 2000).

Photo 1. MSW incineration plants in Amsterdam, Brescia & Vienna respectively

2.2 Typical Incineration plant

A typical incineration plant includes:

Weighing System

The system for weighing solid waste aims at the control and recording of the incoming loads and it has to be practical so as to minimize the time that vehicles remain at this point.

Reception Site

Due to the fact that waste does not arrive on continuous basis (contrary to the feeding of the facility), the existence of waste reception and temporary storage site is considered necessary. The design of the site is made in a way that the following are ensured:

- the unloading time is as little as possible
- all transferred waste is received
- the homogeneity of the waste that will be used as feeding material is achieved
- the smooth feeding of the facility is ensured

Moreover, the design of the reception site should be based on the minimization of the environmental consequences. For instance, the solid waste should remain for maximum two days so as to avoid odours, while the bottom of the site has to be characterised by weathering to allow the leachates and washing wastewater to go away.

Feeding System

The feeding system has to be adapted to the rate and feeding velocity of the installation.

Combustion Hearths

The ignition of solid waste at incineration facilities is achieved through the use of specific burner, which operates with secondary fuel. Basic parameters for the appropriate operation of the combustion hearths are:

- achievement of the minimum desired temperature
- adequate combustion time
- achievement of turbulence conditions / homogenous waste incineration

Boiler

The boiler is the system with which the energy content of the fuel material (hot off-gases) can be utilized in a suitable way through steam production (e.g. at neighbouring industrial facilities or for the heating of urban areas. Pressure, temperature and steam production rate are basic parameters for the effective operation of the boiler.

System for the removal of residues

Residues represent 20 - 40% of the weight of the initial waste and are categorized into:

- Residues that go out of the grates: 20 - 35%
- Residues that go through the grates: 1 - 2%

The residues are collected at hoppers where they are transferred with specific system for cooling.

Emission control system

The role of the emission system control focuses on particles, HCl, HF, SO_2, dioxins and heavy metals and is discussed below (Niessen, 2002).

2.3 Typical emissions from incinerations

The emissions derived from the operation of typical incinerations plants include:

1. Air emissions

The generated air emissions contain the typical combustion products (CO, CO_2, NO_x, SO_2), excess of oxygen, dust particles as well as other compounds. The presence and the concentration of other compounds, such as HCl, HF, suspended particles which contain heavy metals, dioxins and furans, depend on the composition of the waste that is subjected to incineration. During incineration, a quantity of 4,000 - 5,000 m^3 of air emissions is generated per ton of waste.

Air emissions must be controlled by applying appropriate anti-pollution systems, such as:

- Bagfilters
- Electrostatic filters
- Cyclones
- Wet cleaning systems, e.g. scrubbers, wet cleaning towers, rotate sprayers, etc.

Dioxin or furan refers to molecules or compounds composed of carbon and oxygen. These compounds when reacting with halogens, such as chlorine or bromine, acquire toxic properties. Most research on halogenated dioxin and furan has been concerned with chlorinated species. It is generally accepted that dioxin and furan are by-products of combustion processes including domestic and medical waste combustion or incineration processes. In combustion processes, hydrocarbon precursors react with chlorinated compounds or molecules to form furans or/and dioxins. They may also be generated in a post-combustion flue gas cooling system due to the presence of precursor compounds, free chlorine, or unburned carbon and copper species in the fly ash particles.

The toxic influence of dioxins and furans had not been made clear until the end of the decade of 80. The application of the MACT Regulations led to the drastic reduction of the TEQ-Toxic equivalent of the dioxin emissions. As a result, the dioxin emissions have been limited to one thousandth in relation to the year 1987, reaching values lower than 10 gr TEQ per year (Fig. 4). It should also be noted that on the basis of data provided by the US EPA the uncontrolled burning of waste is considered as the main source of dioxins, producing around 600 gr annually.

Dioxins and furans are produced in almost all combustion processes, in the gas phase, while the exact mechanism for their formation is not known. It is known that their formation temperature is 300C°, temperatures where two reactions are possible, formation and decomposition. The existence of chlorinated organic substances in waste and the increase of their content in oxygen encourage their formation. Consequently, the operating conditions of incinerators influence the dioxin formation at higher degree than the waste composition and PVC quantity included in it.

There are indications for the contribution of dioxins and furans to human cancers, fact that makes necessary to take basic and secondary measures so as to limit such emissions.

In order to remove the suspended particles and the gas pollutants, different cleaning systems can be applied. Indicatively, deposition chambers, where 40% of suspended solids is removed, cyclones (removal efficiency 60-80%), wet cleaning towers (removal efficiency 80-95%), electrostatic precipitators (removal efficiency 99-99.5%) and bagfilters (removal efficiency 99.9%) are referred. Apart from the removal of suspended solids, it is often necessary to remove other gas pollutants in the case that they exceed the limit values (like HCl generated during the combustion of PVC and oxides of nitrogen, sulphur and phosphorus. Next, the main cleaning systems that are used for the treatment of the gas products during incineration are briefly described.

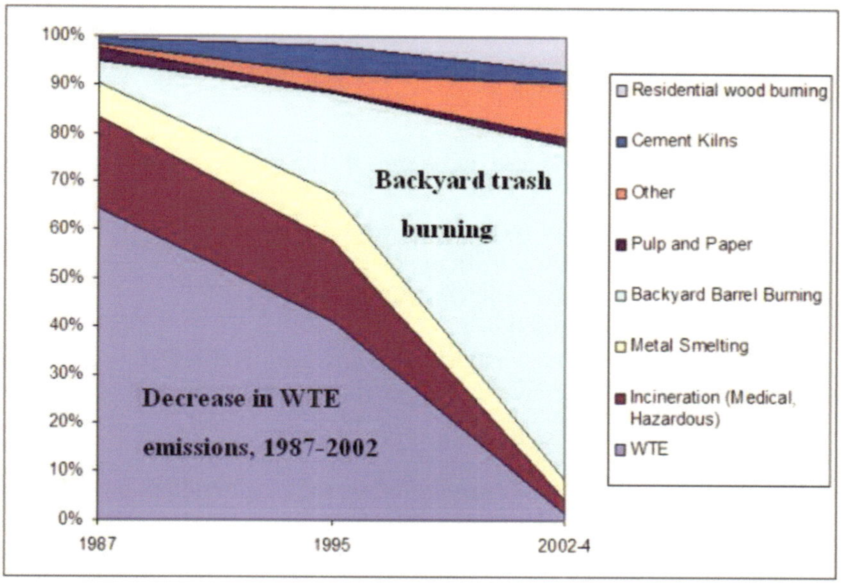

Fig. 4. Dioxins emission in the USA (Deriziotis, 2004).

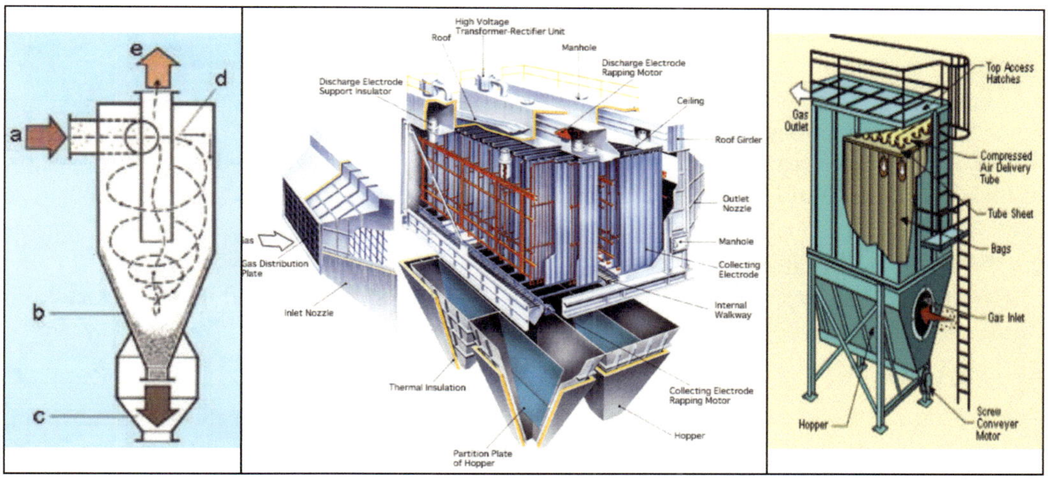

Fig. 5. Cyclones (left), electrostatic precipitators (middle) & bagfilters (right)

Bagfilters: The gases go through porous materials, where the suspended particles are detained. Depending on the requirements, the material of the filters is from natural fibres, plastic ones, glass, minerals, etc. Dust that is collected at the filter cells is removed by vibration or knocks or contrary air provision.

Electrostatic Precipitators (Electro filters): They are consisted of the cathode that can be a simple thin wire and the anode. Another configuration includes a system of parallel tablets, with potential difference between them. Voltage with values between 30-80 KV is developed between the anode and the cathode. When the particles enter the cathode field, they are charged and the negative ones are moving to the positive pole (anode). The velocity of the particles depends on the weight and the Coulomb forces that are developed.

Cyclones: They are based on the development of centrifugal force at the entry of gases at a symmetrical area. The particles due to the centrifugal force and the rotary flow are led towards the walls and then moved downward. Cyclones are often applied together with electrostatic precipitators (Allsopp et al., 2001).

2. Wastewater

Wastewater is generated by the use of water during the incineration process and in particular:

- extinguishing of ash (0.1 m^3 of water/tn of waste)
- cooling of air gasses (2 m^3 of water/tn of waste)
- wet absorbance towers (2 m^3 of water/tn of waste)
- electrostatic filters (precipitators)

The wastewater stream contains suspended solids as well as dissolved organic and inorganic substances. It is characterized as hazardous wastewater and specific treatment is required prior to its final disposal.

3. Solid residues

The secondary solid residues that are generated during incineration can be categorized as follows:

- Fly ash: It is the lightest fraction of the generated solid residues and is collected by the appropriate filters (bagfilters or electrostatic filters). The fly ash contains high concentrations of heavy metals and is characterized as hazardous waste stream.
- Bottom ash: It is the residue of the incineration process (inorganic matter) and is collected at the bottom of the incinerator
- Boiler ash
- Filter dust
- Other solid residues generated during the air emissions cleaning

The solid residues stream must be treated prior to their final disposal, while a main portion of their quantities could be recycled by applying specific processes.

2.4 Types of incinerators

There are various types of incinerators, such as moving grate, fixed grate, rotary-kiln, fluidized bed, etc (Fig. 6).

Moving grate

The typical incineration plant for domestic solid waste is a moving grate incinerator. The moving grate enables the movement of waste through the combustion chamber to be optimized to allow more efficient and complete combustion. A single moving grate boiler can handle up to 35 tones of waste per hour, and can operate 8,000 hours per year with only one scheduled stop for inspection and maintenance of about one month's duration. Moving grate incinerators are sometimes referred to as Municipal Solid Waste Incinerators.

The waste is introduced by a waste crane through the "throat" at one end of the grate, from where it moves down over the descending grate to the ash pit in the other end. Here the ash is removed through a water lock.

Part of the combustion air (primary combustion air) is supplied through the grate from below. This air flow also has the purpose of cooling the grate itself. Cooling is important for the mechanical strength of the grate, and many moving grates are also water cooled internally.

Secondary combustion air is supplied into the boiler at high speed through nozzles over the grate. It facilitates complete combustion of the flue gases by introducing turbulence for better mixing and by ensuring a surplus of oxygen. In multiple/stepped hearth incinerators, the secondary combustion air is introduced in a separate chamber downstream the primary combustion chamber.

According to the European Waste Incineration Directive, incineration plants must be designed to ensure that the flue gases reach a temperature of at least 850 °C for 2 seconds in order to ensure proper breakdown of organic toxins. In order to comply with this at all times, it is required to install backup auxiliary burners (often fueled by oil), which are fired into the boiler in case the heating value of the waste becomes too low to reach this temperature alone.

The flue gases are then cooled in the superheaters, where the heat is transferred to steam, heating the steam to typically 400 °C at a pressure of 40 bar for the electricity generation in the turbine. At this point, the flue gas has a temperature of around 200 °C, and is passed to the flue gas cleaning system.

Often incineration plants consist of several separate 'boiler lines' (boilers and flue gas treatment plants), so that waste receival can continue at one boiler line, while the others are subject to revision.

Fixed grate

The older and simpler type of incinerator was a brick-lined cell with a fixed metal grate over a lower ash pit, with one opening in the top or side for loading and another opening in the side for removing incombustible solids called clinkers.

Rotary-kiln

The rotary kiln incinerator is applied by municipalities and by large industrial plants. This type of incinerator has two chambers, a primary chamber and secondary chamber. The primary chamber consists of an inclined refractory lined cylindrical tube. Movement of the cylinder on its axis facilitates movement of waste. In the primary chamber, there is conversion of solid fraction to gases, through volatilization, destructive distillation and partial combustion reactions. The secondary chamber is necessary to complete gas phase combustion reactions.

The clinkers spill out at the end of the cylinder. A tall flue gas stack, fan, or steam jet supplies the needed draft. Ash drops through the grate, but many particles are carried along with the hot gases. The particles and any combustible gases may be combusted in an "afterburner".

Fluidized bed

According to the technology that is applied for this type of incinerator, a strong airflow is forced through a sand bed. The air seeps through the sand until a point is reached where the sand particles separate to let the air through and mixing and churning occurs, thus a fluidized bed is created and fuel and waste can now be introduced (European Commission, 2006).

3. Gasification

3.1 General

Gasification is the thermal process that converts carbon-containing materials, such as coal, petcoke, biomass, sludge, domestic solid waste to syngas which can then be used to produce electric power, valuable products, such as chemicals, fertilizers, substitute natural gas,

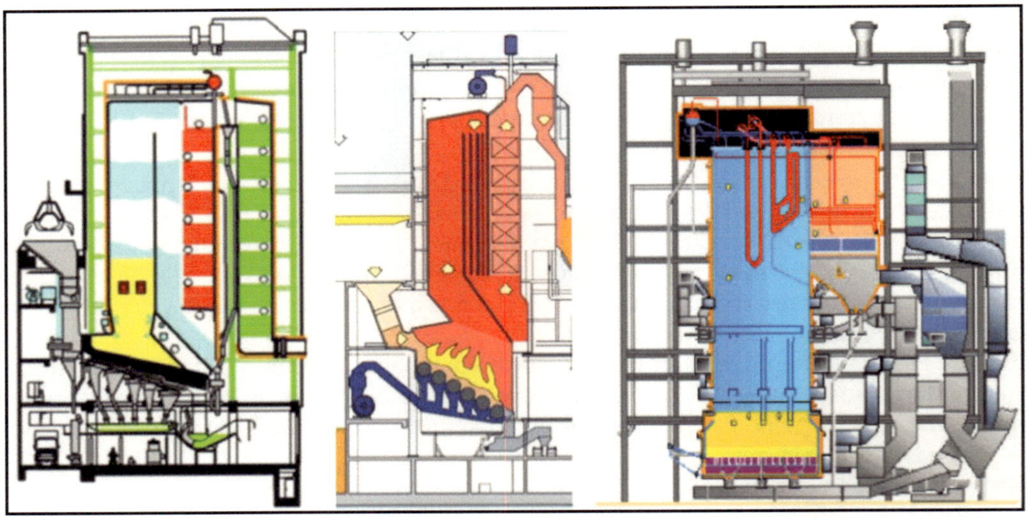

Fig. 6. three types of incinerators: (a) fixed grate, (b) rotary kiln, (c) fluidized bed

hydrogen, steam and transportation fuels. Gasification is defined as a thermal reaction with insufficient oxygen present for reaction of all hydrocarbons (compounds of carbon, hydrogen and oxygen molecules) to CO_2 and H_2O. This is a partial oxidation process which produces a composite gas (syngas) comprised primarily of hydrogen (H_2) and carbon monoxide (CO).

The major benefit of gasification of biowaste is that the product gas can be used directly, after significant cleaning, to fuel a gas turbine generator which itself will form part of a Combined Heat and Power (CHP) or Combined-Cycle Gas Turbine system, thus theoretically improving the overall thermal efficiency of the plant. The main disadvantage is that there can be more items of large equipment and the capital investment is correspondingly higher (Yassin et al., 2009).

The main reactions taking place during gasification are:

$$\text{Oxidation (exothermic) } C + O_2 \rightarrow CO_2 \tag{4}$$

$$\text{Water evaporation reaction (endothermic) } C + H_2O \rightarrow CO + H_2 \tag{5}$$

$$CO + H_2O \rightarrow CO_2 + H_2 \text{ (exothermic)} \tag{6}$$

$$\text{Boudouard Reaction (endothermic) } C + CO_2 \rightarrow 2CO \tag{7}$$

$$CH_4 \text{ formation reaction (exothermic) } C + 2H_2 \rightarrow CH_4 \tag{8}$$

3.2 Typical gasification plant

A typical gasification plant includes:

A) **Feedstock**

Gasification enables the capture – in an environmentally beneficial manner – of the remaining "value" present in a variety of low-grade hydrocarbon materials ("feedstocks") that would otherwise have minimal or even negative economic value. Gasifiers can be designed to run on a single material or a blend of feedstocks:

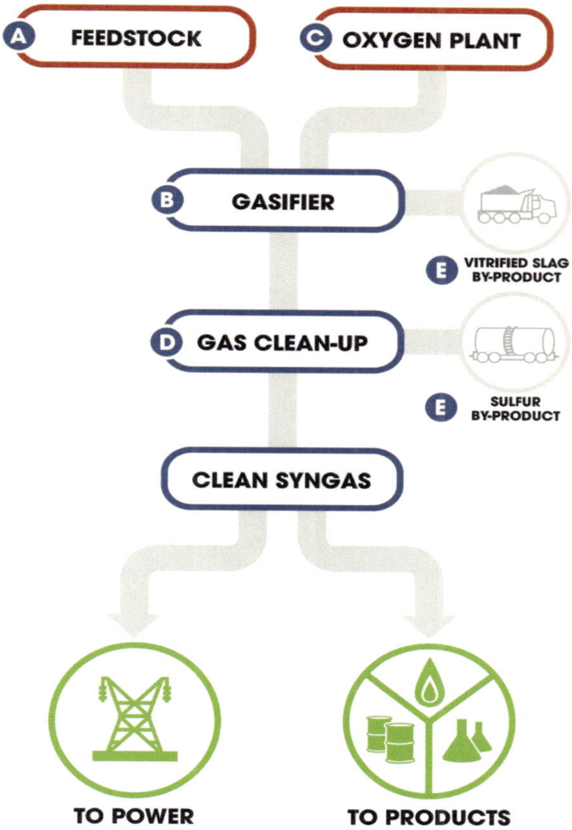

Fig. 7. A schematic diagram of gasification process

- Solids: All types of coal and petroleum coke (a low value byproduct of refining) and biomass, such as wood waste, agricultural waste and household waste
- Liquids: Liquid refinery residuals (including asphalts, bitumen, and other oil sands residues) and liquid waste from chemical plants and refineries
- Gas: Natural gas or refinery/chemical off-gas.

B) <u>Gasifier</u>

The core of the gasification system is the gasifier, a pressurized vessel where the feed material reacts with oxygen (or air) and steam at high temperatures. There are several basic gasifier designs, distinguished by the use of wet or dry feed, the use of air or oxygen, the reactor's flow direction (up-flow, downflow, or circulating), and the gas cooling process. Currently, gasifiers are capable of handling up to 3,000 tons/day of feedstock throughput and this will increase in the near future. After being ground into very small particles – or fed directly (in the case of gas or liquid) – the feedstock is injected into the gasifier, along with a controlled amount of air or oxygen and steam. Temperatures in a gasifier range from 1,400-2,800 degrees Fahrenheit. The heat and pressure inside the gasifier break apart the chemical bonds of the feedstock, forming syngas. The syngas consists primarily of H_2 and CO and, depending upon the specific gasification technology, smaller quantities of CH_4, CO_2, H_2S, and water vapour. Syngas can be combusted to produce electric power and steam or used as a building block for a variety of chemicals and fuels. Syngas generally has a heating value of 250-300 Btu/scf, compared to natural gas at approximately 1,000 BTU/scf. Typically, 70–85% of the carbon in the feedstock is converted into the syngas. The ratio of carbon monoxide to hydrogen depends in part upon the hydrogen and carbon content of the feedstock and the type of gasifier used.

C) Oxygen plant

Most gasification systems use almost pure oxygen (as opposed to air) to help facilitate the reaction in the gasifier. This oxygen (95–99% purity) is generated in a plant using proven cryogenic technology. The oxygen is then fed into the gasifier through separate co-feed ports in the feed injector.

D) Gas Clean-Up

The raw syngas produced in the gasifier contains trace levels of impurities that must be removed prior to its ultimate use. After the gas is cooled, the trace minerals, particulates, sulphur, mercury, and unconverted carbon are removed at high degree using commercially proven cleaning processes common to the chemical and refining industries.
For feeds (such as coal) containing mercury, more than 95% of the mercury can be removed from the syngas using relatively small and commercially available activated carbon beds.

E) By-products

Most solid and liquid feed gasifiers produce a glass-like by-product called slag, which is non-hazardous and can be used in roadbed construction or as roofing material. Also, in most gasification plants, more than 99% of the sulphur is removed and recovered either as elemental sulphur or sulphuric acid.
Hydrogen and carbon monoxide, the major components of syngas, are the basic building blocks of a number of other products, such as chemicals and fertilizers. In addition, a gasification plant can be designed to produce more than one product at a time (co-production or "polygeneration"), such as the production of electricity, steam, and chemicals (e.g. methanol or ammonia). This polygeneration flexibility allows a facility to increase its efficiency and improve the economics of its operations.

3.3 Types of gasifiers

The basic types of the gasifiers are:

- Vertical steady bed
- Horizontal steady bed
- Fluidized bed (Groi et al., 2008)
- Multiple hearths
- Rotary kiln

Among the total five types of installations, the development of vertical and horizontal steady bed facilities, as well as fluidized bed ones is more common.
The facilities of vertical steady bed have advantages, such as the fact that they are simple and have low capital cost, but they are influenced directly by the variations in the composition of the incoming waste (it has to be homogenous, e.g. RDF in condensed form - pellets).
On the basis of the results of pilot applications for units that were operating at temperatures from 650 to 820°C, it was proved that:

- The produced solid residue has high absorption ability and can be used in facilities for the tertiary treatment of water and wastewater.
- The gas product can be used as fuel in oil combustion engines in a proportion 4:1, with the engine performance reaching 76% of the performance in the case that only oil was used (Belgiorno et al., 2003).

Photo 2. MSW gasification plant in Chiba (Japan)

Summarizing, gasification is not an incineration or combustion process. If gasification was compared with incineration, it could be supported that gasification is a conversion process that produces more valuable and useful products from carbonaceous material. Both gasification and combustion processes convert carbonaceous material to gases. Gasification processes operate in the absence of oxygen or with a limited amount of oxygen, while combustion processes operate with excess oxygen.

The objectives of combustion are to thermally destruct the feed material and to generate heat. In contrast, the objective of gasification is to convert the feed material into more valuable, environmentally friendly intermediate products that can be used for a variety of purposes including chemical, fuel, and energy production. Elements generally found in a carbonaceous material such as C, H, N, O, S, and Cl are converted to a syngas consisting of CO, H_2, H_2O, CO_2, NH_3, N_2, CH_4, H_2S, HCl, COS, HCN, elemental carbon, and traces of heavier hydrocarbon gases. The products of combustion processes are CO_2, H_2O, SO_2, NO, NO_2 and HCl.

From an environmental standpoint, gasification offers several advantages over the combustion of solids, heavy oils, and carbonaceous industrial and domestic waste. First, emission of sulphur and nitrogen oxides precursors to acid rain, as well as particulates from gasification are reduced significantly due to the cleanup of syngas. Sulphur in the gasifier feed is converted to H_2S, while nitrogen in the feed is converted to diatomic nitrogen (N_2) and NH_3. Both H_2S and NH_3 are removed in downstream processes, producing a clean syngas. Therefore, if the resulting clean syngas is combusted in a gas turbine to generate electricity or in a boiler to produce steam or hot water, the production of sulphur and nitrogen oxides are reduced significantly. If the clean syngas is used as an intermediate product for manufacture of chemicals, these acid-rain precursors are not formed.

The particulates in the raw syngas are also significantly reduced due to multiple gas cleanup systems used to meet gas turbine manufacturers' specifications. Particulate removal takes place in primary cyclones, scrubbers, or dry filters and then in gas cooling and acid gas removal systems.

A second major advantage is that furan and dioxin compounds are not formed during gasification. Combustion of organic matter is a major source of these highly toxic and carcinogenic pollutants. The reasons why furans and dioxins are not formed in gasification are:

1. The lack of oxygen in the reducing environment of the gasifier prevents formation of free chlorine from HCl and limits chlorination of any precursor compounds in the gasifier
2. High temperature of gasification processes effectively destroys any furan or dioxin precursors in the feed

Furthermore, if the syngas is combusted in a gas turbine where excess oxygen is present, the high combustion temperature does not favor formation of free chlorine. In addition, post-combustion formation of dioxin or furan is not expected to occur because very little of the particulates that are required for post-combustion formation of these compounds are present in the flue gas.

Limited data is available on the concentration of volatile organic compounds, semi-volatile organic compounds (SVOCs), and polycyclic aromatic hydrocarbons (PAHs) from gasification processes. The data that is available indicate that VOCs, SVOCs, and PAHs are either non-detectable in flue gas streams from IGCC process or, in some cases where they were detected, they are at extremely low levels (on the order of parts per billion and lower). The analysis of syngas also indicates greater than 99.99 percent chlorobenzene and hexachlorobenzene destruction and removal efficiencies and part per billion or less concentration of selected PAHs and VOCs (Rezaiyan & Cheremisinoff, 2005; Klein, 2002; Radian International LLC, 2000).

4. Pyrolysis

Pyrolysis is the thermal degradation of carbon-based materials through the use of an indirect, external source of heat, typically at temperatures of 450 to 750°C, in the absence or almost complete absence of free oxygen. This drives off the volatile portions of the organic materials, resulting in a syngas composed primarily of H_2, CO, CO_2, CH_4 and complex hydrocarbons. The syngas can be utilized in boilers, gas turbines or internal combustion engines to generate electricity. The balance of the organic materials that are not volatile are left as char material. Inorganic materials form bottom ash that requires disposal, although some pyrolysis ash can be used for manufacturing brick materials. Pyrolysis involves the thermal degradation of organic waste in the absence of free oxygen to produce a carbonaceous char, oils and combustible gases.

Although pyrolysis is an age-old technology, its application to biomass and waste materials is a relatively recent development. An alternative term for pyrolysis is thermolysis, which is technically more accurate for biomass energy processes because these systems are usually starved-air rather than the total absence of oxygen. Although all the products of pyrolysis may be useful, the main fuel for power generation is the pyrolysis oil. Depending on the process, this oil may be used as liquid fuel for burning in a boiler or as a substitute for diesel fuel in reciprocating engines, although this normally requires further processing (Institution of Mechanical Engineers, 2007).

The reactions taking place initially are decomposition ones, where organic components of low volatility are converted into other more volatile ones:

$$CxHy \rightarrow CcHd + CmHn \tag{9}$$

Moreover, at the early stages of pyrolysis process, reactions occurring include condensation, hydrogen removal and ring formation reactions that lead to the formation of solid residue from organic substances of low volatility:

$$CxHy \rightarrow CpHq + H_2 + coke \tag{10}$$

In the case of existence of oxygen, CO and CO_2 are produced or the interaction with water is possible. The produced coke can be vaporized into O_2 and CO_2.
The pyrolysis products can be liquid, solid and gaseous. The majority of the organic substances in waste are subjected to pyrolysis by 75 - 90 % into volatile substances and by 10 - 25 % to solid residue (coke). However, due to the existence of humidity and inorganic substances, the quantity of volatile substances varies from 60 to 70% and the coke between 30 and 40%.
In order to achieve the successful operation of a pyrolysis facility, continuous control is required due to the complex processes taking place during the method development. Moreover, solid waste with no major composition variation that does not include metals and glass has to be fed on continuous basis (use of waste after successful implementation of separation at source or mechanical separation). In addition, special care is needed about

Solid	Carbon that is incorporated into several inert products	-
Gas	Dust particles, CO, CO_2, CH_4, H_2	700 m^3 off-gases / tone of waste
Liquid	CH_3COOH, CH_3COCH_3, CH_3OH, complex oxygenised H/C	

Table 1. Brief description of the solid, liquid and gas products from the operation of a pyrolysis unit

% v/v gas composition	Pyrolysis temperature (°C)			
	500	650	815	926
CO	33.6	30.5	34.1	35.3
CO_2	44.8	31.8	20.6	18.3
H_2	5.6	16.5	28.6	32.4
CH_4	12.5	15.9	28.6	32.4
C_2H_6	3.0	3.1	0.8	1.1
C_2H_4	0.5	2.2	2.2	2.4
Calorific Value (btu/St/t)	312	403	392	385

Table 2. Composition of the produced gas at different pyrolysis temperatures

whether the liquid products satisfy the specifications of commercial fuel (mainly due to the humidity within these products).
The product proportions depend on the waste nature, the temperature conditions and the treatment time.
The products produced from pyrolysing materials are a solid residue and a synthetic gas (syngas), while some of the volatile components form tars and oils can be removed and reused. The solid residue (sometimes described as a char) is a combination of non-combustible materials and carbon. The syngas is a mixture of gases (combustible constituents include carbon monoxide, hydrogen, methane and a broad range of other VOCs). A proportion of these can be condensed to produce oils, waxes and tars. The syngas typically has a net calorific value of between 10 and 20 MJ/Nm^3. If required, the condensable fraction can be collected by cooling the syngas, potentially for use as liquid fuel (Gidarakos 2006).

Typical Pyrolysis Facility

In a typical pyrolysis facility the following are taking place:

- Drying of solid waste (100-200°C)
- Initial decomposition of substances, initiation of the decomposition of H_2S and CO_2 (250°C)
- Break of the bonds of aliphatic substances - Start of the separation of CH_4 and other aliphatic substances (340°C).
- Enrichment of the produced material in carbon (380°C)
- Break of the bonds C - O and C - N (400°C).
- Conversion of coal tar materials into fuel material and tar (400 - 600°C).
- Decomposition to materials resistant to heat - Formation of aromatic substances (600°C).
- Production of aromatic substances, processes for hydrogen removal from organics like butadiene, etc. (>600°C).

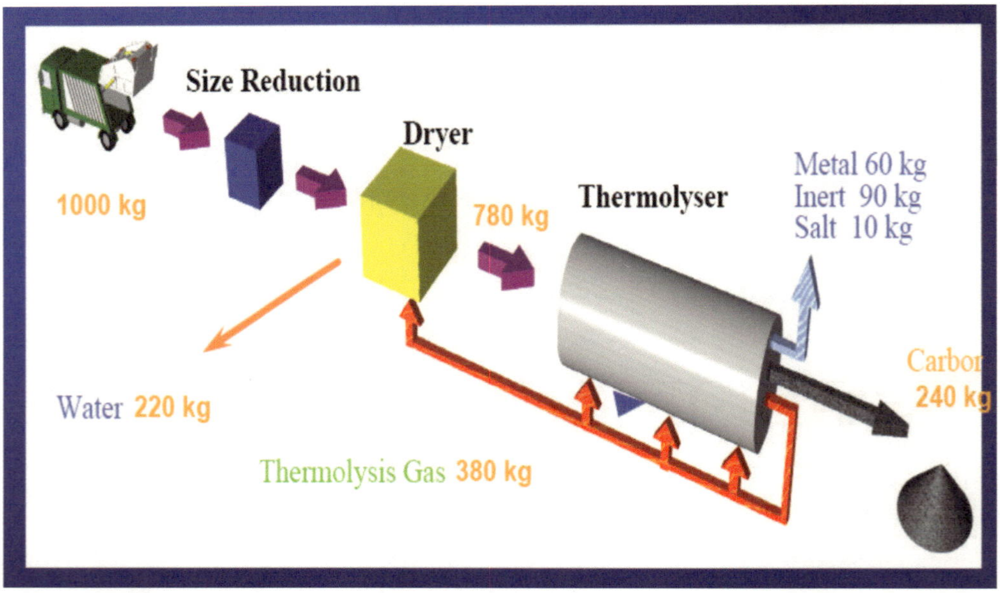

Fig. 8. A schematic diagram of gasification process

The main advantages of pyrolysis in comparison to incineration are:

- The decomposition temperature is lower than the incineration temperature, so the thermal distress of the whole facility is less intense than in incineration.
- The decomposition takes place in reducing atmosphere and not in oxidizing like in incineration. The demand for less oxygen is also the reason for less air emissions in the case of pyrolysis.
- The ash content in carbon is much higher than in the case of incineration.
- The metals that are included in waste are not oxidized during pyrolysis and have higher commercial value.
- The produced gas is at different hearth and probably other site from the pyrolytic reactor.
- No ash is produced from the combustion of the pyrolysis gas and the cleaning of the off-gas is a simpler process.
- The initial waste volume is reduced at higher level in comparison with the incineration.

The main disadvantages of pyrolysis include:

- The big problem of this technique is that pre-treatment is required including cutting and separation of waste prior to pyrolysis which can increase the cost for the installation and operation of such units substantially.
- The pyrolysis products cannot be disposed without further treatment.
- The facilities for cleaning the gases and wastewater require extremely high cost.
- At present, the application of the method at large scale is limited. Nevertheless, the prospects for reactors of average temperature with the form of rotary drum or fluidized bed seem to be better.

5. Plasma technology

5.1 General

Plasma refers to every gas of which at least a percentage of its atoms or molecules is partially or totally ionized. In a plasma state of matter, the free electrons occur at reasonably high concentrations and the charges of electrons are balanced by positive ions. As a result, plasma is quasi-neutral. It is generated from electric discharges, e.g. from the passage of current (continuous, alternate or high frequency) through the gas and from the use of the dissipation of resistive energy in order to make the gas sufficiently hot. Plasma is characterized as the fourth state of matter and differs from the ideal gases, because it is characterized by 'collective phenomena'. 'Collective phenomena' originate from the wide range of Coulomb forces. As a result, the charged particles do not interact only with neighbouring particles through collisions, but they also bear the influence of an average electromagnetic field, which is generated by the rest charges. In a large number of phenomena, collisions do not play important role, as 'collective phenomena' take place much faster than the characteristic collision time (Blahos, 2000).

Plasma technology can be used as a tool for green chemistry and waste management (Mollah et al., 2000). Thermal plasmas have the potential to play an important role in a variety of chemical processes. They are characterized by high electron density and low electron energy. Compared to most gases even at elevated temperatures and pressures, the chemical reactivity and quenching rates that are characteristic of these plasmas is far greater. Plasma technology is very drastic due to the presence of highly reactive atomic and ionic species and the achievement of higher temperatures in comparison with other thermal methods. In fact, the extremely high temperatures (several thousands degrees in Celsius scale) occur only in the core of the plasma, while the temperature decreases substantially in the marginal zones (Leal-Quirós, 2004; Gomez et al., 2009).

Five distinct categories of processes are used as the basis for the plasma systems catering for waste management. These are:

- Plasma pyrolysis (Huang & Tang, 2007; Sheng et al., 2008)
- Plasma combustion (also called plasma incineration or plasma oxidation)
- Plasma vitrification
- Plasma gasification in two different variants (Malkow, 2004)
- Plasma polishing using plasma to clean off-gases

The proportion of air used during waste treatment and the nature of the output products are primary differences between the aforementioned plasma processes. In practice, commercial processes can be designed to allow two or more of these to occur within a single integrated system (Juniper 2006).

Plasma gasification is the most common plasma process. It is an advanced gasification process which is performed in an oxygen-starved environment to decompose organic solid waste into its basic molecular structure. Plasma gasification does not combust the waste as incinerators do. It converts the organic waste into a fuel gas that still contains all the chemical and heat energy from the waste. Also, it converts the inorganic waste into an inert vitrified glass.

Electricity is fed to a torch, which has two electrodes, creating an arc. Inert gas is passed through the arc, heating the process gas to internal temperatures as high as 25,000 degrees Fahrenheit. The following diagram illustrates how the plasma torch operates.

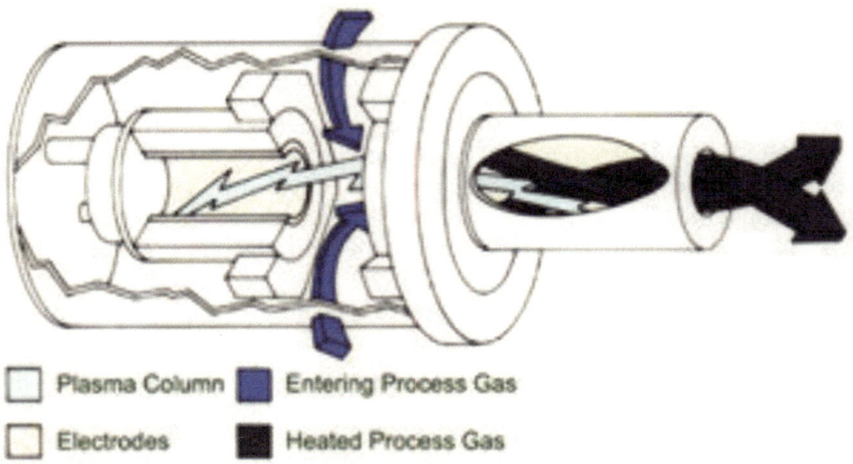

Fig. 9. Illustration of the operation of a plasma torch (Westinghouse)

The temperature a few feet from the torch can be as high as 5,000-8,000°F. Because of these high temperatures waste is completely destroyed and broken down into its basic elemental components. At these high temperatures all metals become molten and flow out the bottom of the reactor. Inorganics, such as silica, soil, concrete, glass, gravel, etc. are vitrified into glass and flow out the bottom of the reactor. There is no ash remaining to go back to a landfill.

Mixed solid waste is shredded and fed into a reactor where an electric discharge similar to a lightning (the plasma) converts the organic fraction into synthesis gas and the inorganic fraction into molten slag. Typically temperatures are greater than 7,000°F achieving complete conversion of carbon-based materials, including tars, oils, and char, to syngas composed primarily of H_2 and CO, while the inorganic materials are converted to a solid, vitreous slag. The syngas can be utilized in boilers, gas turbines, or internal combustion engines to generate electricity, while the slag is inert and can be used as gravel.

Disadvantages of the process include: Relatively high cost, high level of maintenance and, skilled labour required for operations.

The plasma technology can be used for the thermal treatment of any type of waste. The only variable is the amount of energy that it takes to destroy the waste. Consequently, no sorting of waste is necessary and any type of waste, except nuclear waste, can be processed.

The plasma reactor operates at a slightly negative pressure, meaning that the feed system is simplified because the gas does not want to escape. The gas has to be pulled from the reactor by the suction of the compressor. Because of the size and the negative pressure, the feed system can handle bundles of material up to 1 meter in size. This means that sizeable waste can be fed directly into the reactor and pre-processing of the waste is not needed. Also, the performance of the plasma gasifier is not affected by the moisture of the waste (during incineration, the moisture of waste consumes energy to vaporize and can impact the capacity and economics of the process).

An indicative list of attempts to use plasma technology in waste treatment applications is as follows:

Location	Waste	Capacity (TPD)	Start Date
Mihama-Mikata, JP	MSW/WWTP Sludge	28	2002
Utashinai, JP	MSW/ASR	300	2002
Kinuura, JP	MSW Ash	50	1995
Kakogawa, JP	MSW Ash	30	2003
Shimonoseki, JP	MSW Ash	41	2002
Imizu, JP	MSW Ash	12	2002
Maizuru, JP	MSW Ash	6	2003
Iizuka, JP	Industrial	10	2004
Osaka, JP	PCBs	4	2006
Taipei, TW	Medical & Batteries	4	2005
Bordeaux, FR	MSW ash	10	1998
Morcenx, FR	Asbestos	22	2001
Bergen, NO	Tannery	15	2001
Landskrona, SW	Fly ash	200	1983
Jonquiere, Canada	Aluminum dross	50	1991
Ottawa, Canada	MSW	85	2007 (demonstration)
Anniston, AL	Catalytic converters	24	1985
Honolulu, HI	Medical	1	2001
Hawthorne, NV	Munitions	10	2006
Alpoca, WV	Ammunition	10	2003

Table 3. Commercial Plasma Waste Processing Facilities (Circeo, 2007)

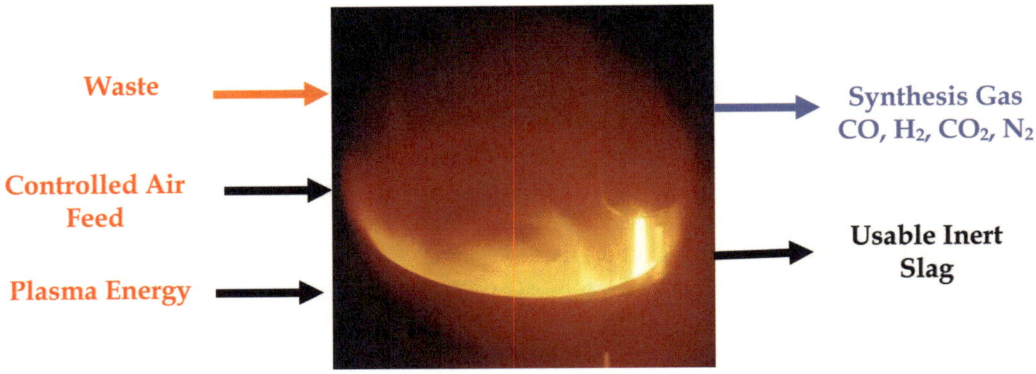

Fig. 10. A schematic diagram of plasma process

There are two generic configurations of plasma gasification: configurations in which the plasma generator is external to the main waste conversion reactor and is used as a source of hot gases (this is often referred to as "plasma assisted gasification") (Fig. 11) and those in which the plasma generator (plasma torch or electrodes) is contained within the main waste conversion reactor (Fig. 12).

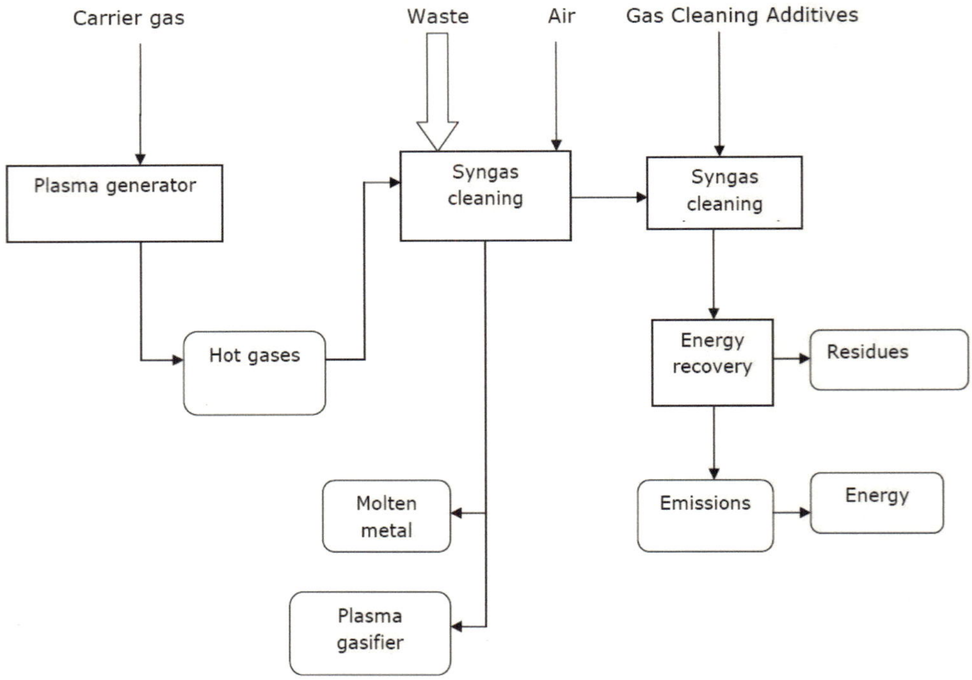

Fig. 11. "Plasma Assisted" gasification (Juniper 2006)

Photo 3. "Plasma Assisted" gasification unit owned by Hitachi Metals with treatment capacity of 200 tones of municipal solid waste and automobile shredder residue per day in Utashinai in Japan (Circeo, 2007)

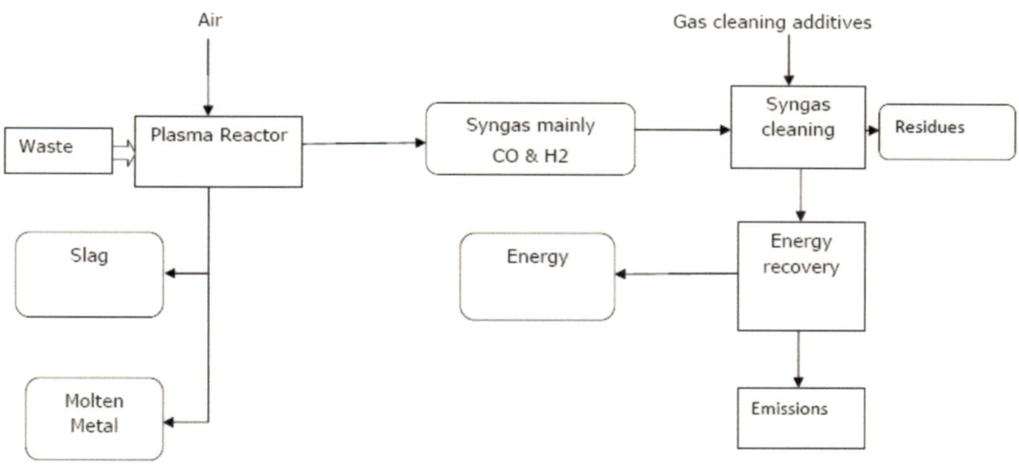

Fig. 12. Plasma gasification (Juniper 2006)

5.2 The case of the pilot gasification / vitrification unit in Greece

The first attempt to apply gasification process in Greece was made by the Unit of Environmental Science & Technology of the National Technical University of Athens, with a unit that was installed in Mykonos in order to treat all types of waste generated on the island. The scope was to investigate the use of this innovative technique in an isolated area like an island in order to provide a solution to the overall management of waste. General views of the whole demonstration facility are available below:

Photo 4. General view of the demonstration gasification / vitrification unit

Photo 5. Another general view of the demonstration gasification / vitrification unit

The primary waste feeding system consists of a hopper intended for feed of solid material having maximum moisture content of 50% and a maximum particle size of 2.5 cm. The screw conveyor solid feeder has a maximum capacity of about 85 kg/h of waste and the

feeding capacity varies depending on the feed waste bulk density. The feed rate is adjustable by varying the speed of the screw conveyor. Waste is manually loaded into the hopper connected to the screw conveyor. The feed rate is continuous and very steady, compared to a hydraulic feeder.
Waste is fed from a hopper through a screw feeder to the top of the furnace and dropping down is passing through the very hot and free of oxygen region between the two electrodes.

Photo 6. Feeding system

The furnace is comprised of a crucible, with approximately 130 litters capacity. It also includes a start-up natural gas burner for preheating and idle operation, a port for gasification air injection, a water-cooling mechanism for the graphite electrodes, an external surface water-cooling for the furnace walls and a tapping hole for periodical or continuous slag removal. During the operation of the plasma unit, the bottom part of the furnace contains the molten slag, while the upper section of it contains the process gases and is lined with a suitable high-temperature refractory. The required gasification air fed to the furnace is supplied by a compressed air system. Adjusting the valves on the compressed air line can control the flow rate.

Photo 7. Gasification / vitrification furnace

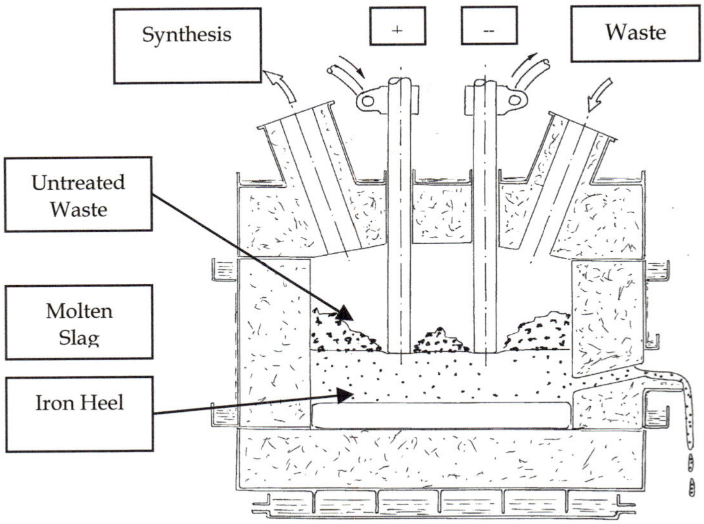

Fig. 13. Plasma Gasification / Vitrification Process

In the pilot unit the furnace in which waste gasification is taking place is preheated at 600-800°C by burning propane in its interior. After preheating, two cylindrical graphite electrodes are inserted in the furnace and their ends are approached to a close distance. Two graphite electrodes are used to supply an electrical arc to the furnace. The current flows from the anode (+) to the molten bath and from the bath to the cathode (-). The cathode is grounded at zero (0) potential.
Graphite electrodes with male/female threads are used. The electrode dimensions were 7.6 cm in diameter and 106.7 in length. Electrodes are installed with the female end down, in

Photo 8. Preheating

order to avoid dust accumulation in the threads. Two electrodes were screwed together on each side (anode and cathode) and are mounted on flexible joints, which allow them to be moved over the slag pool and improve mixing. The mechanism also permits the electrodes' extension into the furnace to be adjusted during operation (Carabin & Holcroft, 2005; Carabin et al., 2004; Gagnon & Carabin, 2006).
The DC power supply for the electrodes has a maximum power output of 200 KVA (Plasma arc power supply, input: 600 VAC-3ϕ-60HZ, 3 X 200A fuses).

Photo 9. Movement Mechanism of the graphite electrode

Photo 10. Camera recording what is happening in the furnace

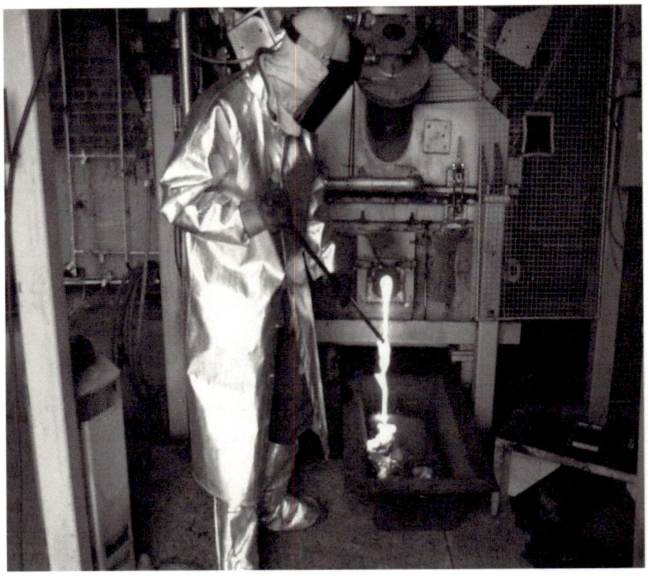

Photo 11. Tapping inert slag (Pyrogenesis Canada Inc, Montreal, Canada)

Then, a high voltage is applied between them producing an electrical arc which is raising locally the temperature up to values as high as 5,000 °C and creating a plasma atmosphere. Air is not permitted to enter the furnace. Under these conditions it is ensured that from the volatile part of the wastes syngas is produced consisting mainly of H_2, CO, CO_2 and H_2O and containing in very low proportions H_2S and HCl, but without significant presence of NOx. A camera is installed in front of a window on the top of the furnace, connected with a laptop, by which we can watch or make video recording of the electrical arc and the decomposition of the organic matter taking place in the interior of the furnace.

Photo 12. No arc

Photo 13. Plasma arc

The slag could be tapped out periodically from the tap hole located on the front side of the crucible, close to the bottom of the furnace. The slag was either poured in a slag mold to form ingots (photo 14) or quenched in a water tank to produce granulated slag (photo 15). The inorganic part of the wastes used is melted, drops to the bottom of the furnace and from time to time is removed through a hole in the lower part of the furnace, is collected to a fire resistant pan and is taken to the laboratory for analysis and investigation of its toxicity.

Photo 14. Air-cooled slag

Photo 15. Water cooled slag in the form of granules

The hot cyclone was designed to remove dust in the synthesis gas. The produced gases while entering the cyclone are put in circular movement and the centrifugal force makes particulate matter contained in the gases to be removed to a high degree.

Photo 16. Cyclone

Photo 17. Secondary Combustion Chamber

The result of the operation of the Secondary Combustion Chamber (SCC) is the oxidation of the components of the furnace gases. It was designed to combust H_2 and CO in the synthesis gas. In order to combust CO and H_2 into CO_2 and H_2O, air is added into the secondary combustion chamber. Propane burners are used to maintain the chamber temperature at

1,100°C. The operator can check local regulations to determine the required temperature in secondary combustion chamber. This temperature is required to fully combust CO and H_2 in a region where no hazardous by-products are created. In normal operation, the gas residence time in the secondary combustion chamber is about two seconds. A single blower provides the combustion air for the burners and the combustion air for the synthesis gas.
The quench vessel is located at the outlet of the secondary combustion chamber. Its role is to cool the combustion gases quickly to approximately 75°C so as to minimize any production of dioxins, furans or other organic compounds. The shock-like cooling avoids the formation of the aforementioned compounds from elementary molecules in the synthesis gas due to the de novo Synthesis back reactions (Calaminus & Stahlberg, 1998). These reactions are known to occur in waste heat boilers where a slow cooling in the range from 400°C to 250°C of flue gases with chlorine compounds, non combusted organic molecules and catalysts such as dust will result in dioxin formation. The quench vessel uses two atomizing nozzles to quench the gas from the secondary combustion chamber. These nozzles are capable of providing 2 litters per minute of flow. Regulating the amount of the quenching water can control the gas temperature exiting the vessel.

Photo 18. Quench Vessel

Photo 19. Scrubber

The scrubber removes water-soluble components of the off-gas including hydrochloric acid and most oxides of sulphur, prior to discharge. Since the synthesis gas may contain acid gases (such as HCl or SO_2), a packed tower type wet scrubber uses caustic soda to neutralize the acid gas from the quench vessel. The pH of the scrubbing solution is controlled at 9.0. The scrubber liquor is re-circulated through a wet bagfilter in order to remove suspended particles. The bagfilter is a cartridge unit having series of cylindrical filters that are cleaned periodically by an automatic sequence using pulses of compressed gas.
To sum up, the pilot unit has a maximum capacity of 50 kg of waste per hour and the quantity of the syngas produced is too low for a gas engine to convert it in electrical energy, therefore the syngas has to be released in the atmosphere but in a safe way. Hence, CO and

H_2 have to be transformed to CO_2 and H_2O and for this purpose the SCC has been added in the installation, which is maintained at high temperature by combusting propane with air and in which CO and H_2 are burnt to CO_2 and H_2O. The SCC in our installation is situated after the furnace and between the two units is interceded a cyclone to remove the solid particles. After the SCC the flue gases are objected to quenching by coming in contact with a big quantity of cold water and this takes place in a pipe where flue gases and cooling water are moving opposite each other. After quenching, the flue gases are passing for cleaning through a scrubber with NaOH solution, then through a filter and finally before they are released to the atmosphere via a stack are cooled in a heat exchanger to condense and recirculate the maximum quantity of water vapors (Moustakas et al., 2005; Moustakas et al., 2008).

The control and monitoring of the operation of the whole system is made through a laptop by a special software program. The main pages of the software are as follows:

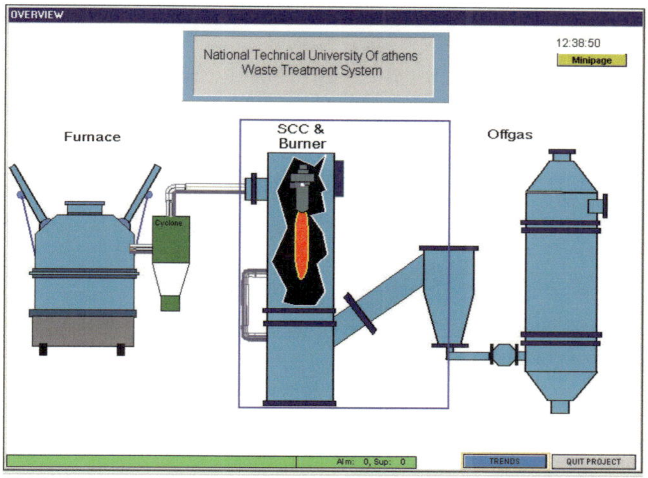

Fig. 14. Main page of the software system for the control of the operation of the unit

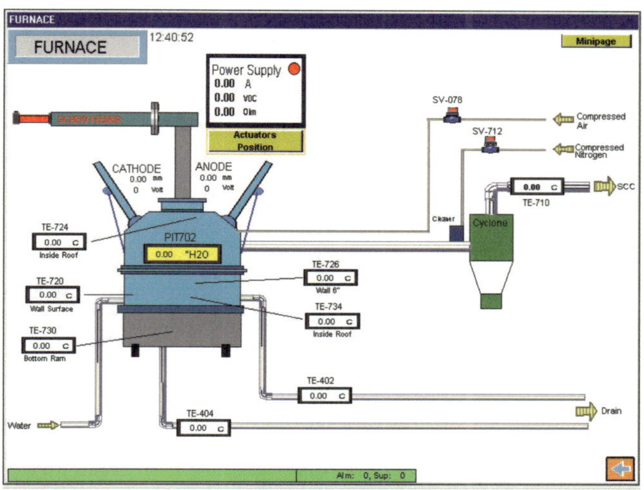

Fig. 15. Page of the software system for the control of the furnace

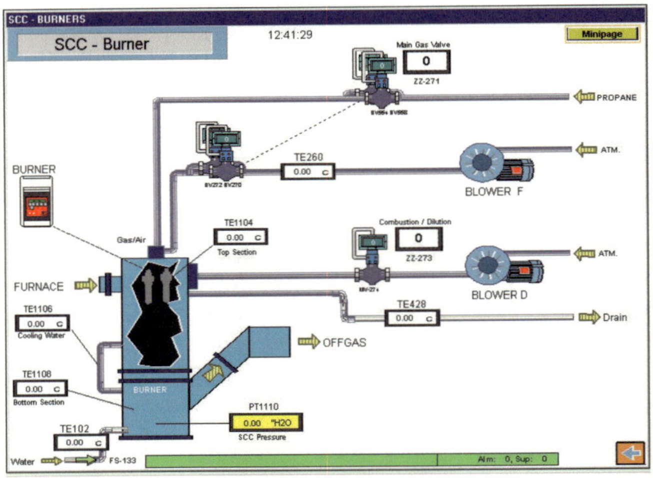

Fig. 16. Page of the software system for the control of the Secondary Combustion Chamber

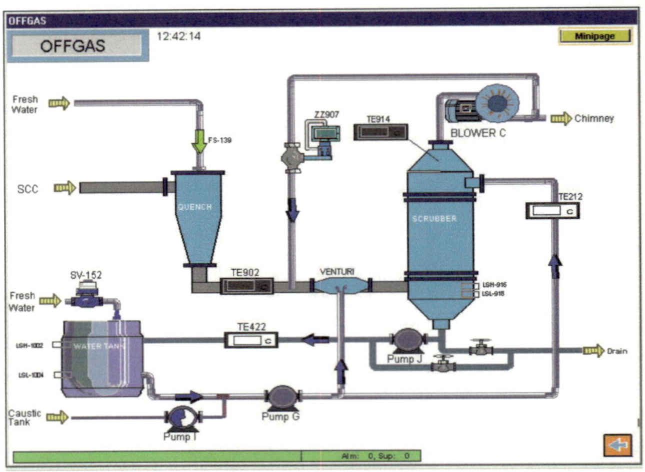

Fig. 17. Page of the software system for the controlof the off-gas cleaning systems

Two gas analyzers for on line measurements are also necessary for the full monitoring of the operation of the gasification / vitrification plant, one between the furnace and the SCC for the determination of the syngas composition and the other one in the outlet stack of the flue gases to determine their composition and avoid air pollution.

The main consumables needed for its operation are water, diesel to run the generator for the power supply of the installation, propane as well as the graphite electrodes.

Regarding the produced slag, after the vitrification, the slag was studied by X-ray fluorescence analysis for the determination of chemical composition, by conventional Bragg-Brentano X-ray diffraction (XRD) for the evaluation of crystalline phases formed and by scanning electron microscope (SEM) equipped with energy-dispersive X-ray (EDX) elemental analysis for microstructure/morphology observation and compositions (Kuo et al., 2009).

The amorphous or crystalline nature of vitreous materials was established by X-ray diffraction (XRD). Samples were crushed to fine powder in an agate mortar and then were scanned with

CuKα radiation from $10 \leq 2\theta \leq 70°$ at a scanning speed of 0.3°/min, using a Siemens D5000 powder X-ray diffraction unit, operating at 30 mA and 40 kV. The XRD analysis patterns are shown in Fig. 18 and 19 for water quenched and air-cooled slag respectively.

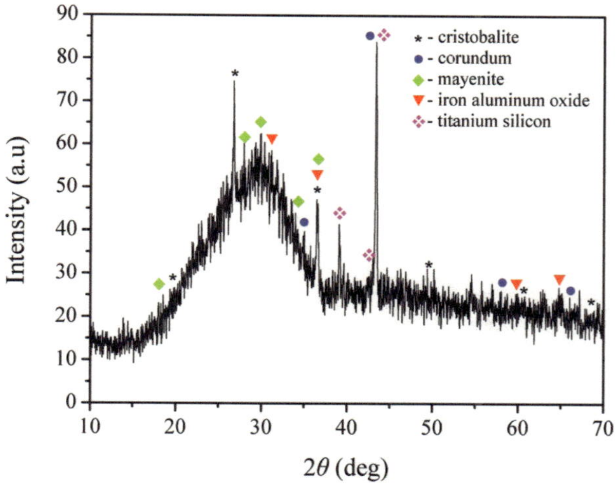

Fig. 18. XRD of the water quenched slag

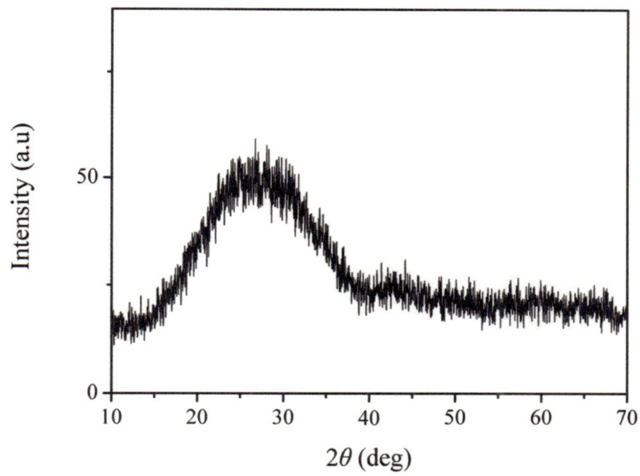

Fig. 19. XRD of the air-cooled slag

The XRD pattern (Fig. 18) indicates that the water quenched slag is composed of mainly amorphous and traces of crystalline phase. Crystalline phases were identified by comparing intensities and positions of Bragg peaks with those listed in the Joint Committee on Powder Diffraction Standards (JCPDS) data files. The crystalline phases that could be identified were cristobalite (SiO_2), corundum (Al_2O_3), mayenite ($Ca_{12}Al_{14}O_{33}$) and iron aluminum oxide ($Fe_{1.006}Al_{1.994}O_4$).

The XRD pattern of the air cooled slag revealed an amorphous phase and no crystalline structures or phases are observed (Fig. 19). The formation of glassy amorphous structures

drastically reduces the specific surface area and present better resistance to the decomposition by an acid than the crystalline structure.
The SEM micrographs in Fig. 20 illustrate the morphology of the two slag types. More specifically, no significant differences were noted and the common conclusion is that both water-cooled and air-cooled slags are characterized as equable.

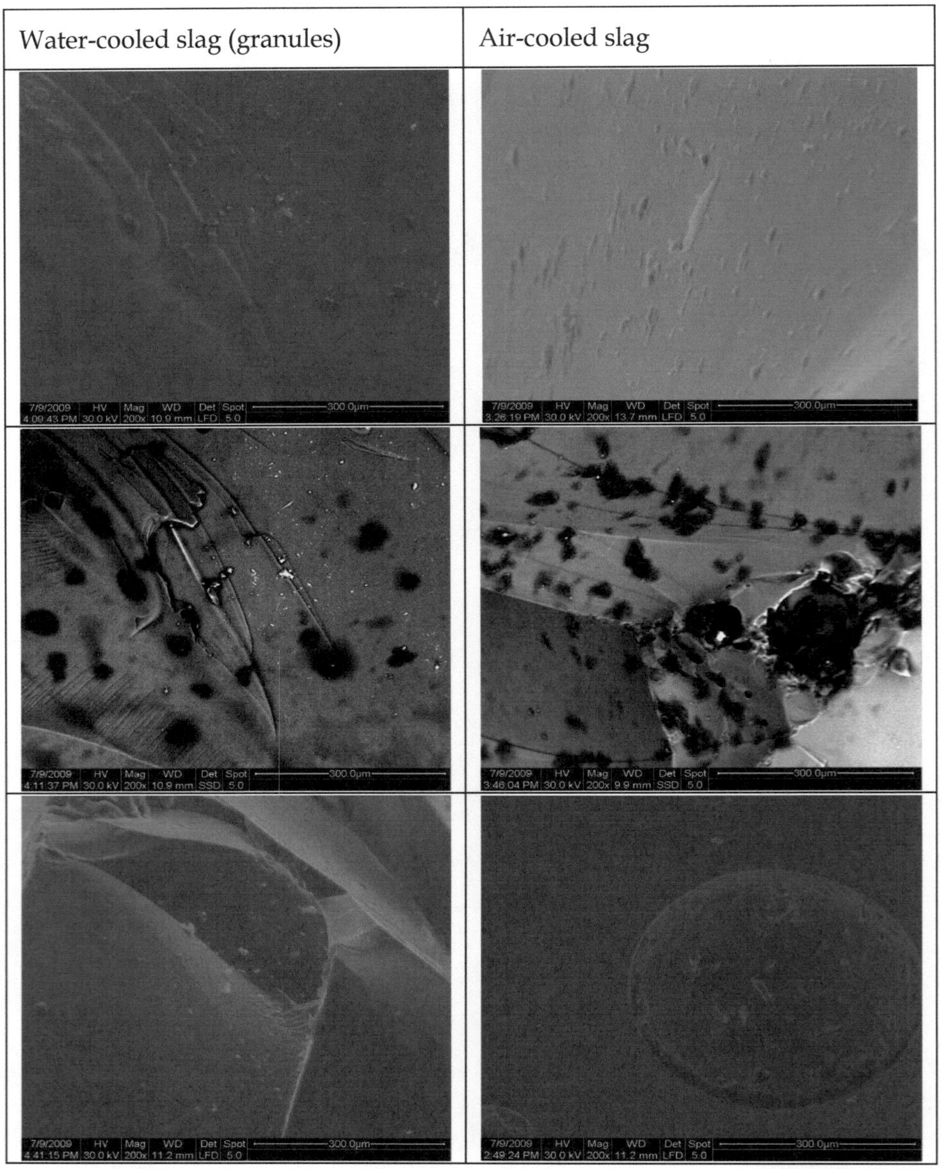

Fig. 20. SEM images

Consequently, the SEM images make us conclude that the slight crystalline areas present in water quenched slag are enclaved and, therefore, both types of solid residues are considered really stable and inert.

On the basis of the primary results derived from the operation of the demonstration gasification facility in Mykonos and elsewhere, plasma gasification is a promising technology especially in the case of isolated areas, such as islands. More specifically,

- The method is characterized by relatively low air emissions that are not harmful for the environment. The release of polluting substances, such as SO_2, metals, dioxins will be at much lower levels than conventional thermal techniques like incineration.
- Gasification can be used for the management of all types of waste, both hazardous and non hazardous waste. Such facilities can handle municipal, toxic and hospital waste or mixtures of them
- Plasma gasification is not an incineration process. As a result, the disadvantages of the incineration are avoided.
- No ash or other by-products, such as biomass that has to be disposed at landfills after the treatment. In this way, there is no disposal cost provided that there is market for the vitrified slag.
- The material recovery is greater than in any other thermal technique. Instead of consuming raw materials, this method produces slag that can be used as material in a variety of applications, such as construction works.
- Energy recovery is higher than any other waste management practice. Therefore, the income for energy sale can be significant. It is supported that in the case of plasma gasification the generation of net electricity (steam turbine power generation) from 1 tone of municipal solid waste could reach the value of 816kWh. The relevant net electricity from pyrolysis (Mitsui R21 Technology) is 571 kWh and 544 kWh from mass-burn technology (Circeo 2007).
- The emissions at air, water and soil are lower than in other processes.
- Plasma gasification can be used for energy production from non gas fuels.
- The releases to the atmosphere during the production of electrical energy are similar with those of facilities with natural gas.
- Since every C-based substance that exists in the plasma gasifier is converted to gas, each of them can be used as fuel (Lemmens et al., 2007).

6. Conclusions

The energy utilization from waste can be achieved with the application of different thermal technologies (anaerobic digestion, a biological waste management method, can also result in energy recovery form waste). The basic operation principles that should apply to all thermal treatment facilities for municipal solid waste are:

1. Steady operation conditions.
2. Easiness for adaptation to rough changes of the composition and the quantity of feedstuff.
3. Flexibility for adaptation to the variations of the composition and the quantity of the used fuel.
4. Full control of the pollutants in the emissions.
5. Maximization of the utilization of the thermal energy, mainly for the production of electrical energy.
6. Minimization of the capital and operation cost.

Summarizing the main characteristics of the common thermal techniques for waste management, the following table presents the basic products and the main operation conditions.

Parameter	Incineration	Pyrolysis	Gasification
Operation conditions			
Temperature °C	800-1,450	250-700	500-1,600
Pressure (bar)	1	1	1-45
Atmosphere	Air	Inert/Nitrogen	Gasification factor: O_2, H_2O
Stoichiometric relation	>1	0	<1
Products			
Gas Phase	CO_2, H_2O, O_2, N_2	H_2, CO, H_2O, N_2, H/C	H_2, CO, CO_2, CH_4, H_2O, N_2
Solid Phase	Ash, Scoria	Ash, Scoria	Ash, Scoria
Liquid Phase		Pyrolysis Oils & H_2O	

Table 4. Parameters of typical operation conditions & products of the common thermal management practices

Thermal waste management methods should be applied together with separation at source of all materials that can be recycled in order to maximize material recovery from waste. The advantages of thermal methods in waste treatment are summarized as follows:

- Reduction of the weight and volume of the treated waste: The final solid residues have weight that varies from 3 to 20% in relation to the initial weight of waste, depending on the technology that is used. Gasification and pyrolysis result in lower quantities of solid residues comparing to incineration.
- Absence of pathogenic factors in the products:
  - The products of thermal treatment, due to the high temperatures that are developed, are characterized from complete absence of pathogenic factors.
- Demand for limited areas:
  - The thermal treatment units are characterized by low demands for land for their installation.
  - The pyrolysis and gasification processes require less space in relation to incineration.
- Utilization of the energy content of waste:
  - Through the thermal treatment technologies, the exploitation of the energy content of waste is possible.
  - This energy can be either electric or thermal energy.
- Reduction of the burden paused to the landfill sites and consequent increase of their lifetime.
- Extraction of the organic fraction of municipal waste from landfill sites, as required by the relevant legislative framework (Directive 1999/31/EC).

Indicative disadvantages of the application of thermal methods are the following:

- Relatively high capital cost:
  - Higher than that of other technologies for the management of municipal waste.
  - Significant part of the total capital cost, especially for the case of incineration, is spent on antipollution measures.
- Increased operation cost

- In general, the thermal management techniques are characterized by relatively high operation cost. The cost is reduced substantially as the capacity of the plant increases.
- Demand for high quantities of waste:
  - Especially for the case of incineration - combustion, a minimum capacity is required so that the units are financially feasible. Estimated minimum served population from incineration facilities is 100,000 inhabitants (around 50,000 tones of waste annually). Gasification and pyrolysis can be applied for much lower waste quantities (around 15,000 tones of waste per year)
- Need for specialized personnel.

Regarding the first pilot application for waste gasification in Greece, an EU country where the thermal management of municipal waste is not applied, the main advantages of the process involve: good environmental performance, production of more than 500 KWh net of electricity per tone of waste treated, no by-products going to landfill. Therefore, it is hoped that this attempt will lead to full scale gasification facility in Mykonos, which will cater for the needs of the whole island treating municipal as well as other waste streams (e.g. hospital waste), with total capacity in the range between 10,000 and 15,000 tones per year. The fulfilment of the whole project will constitute innovative achievement at European level and will be an effective waste management success story for isolated areas and especially islands.

7. References

Allsopp, M., Costner, P. & Johnston, P. (2001). *Incineration and human health, State of knowledge of the impacts of waste incinerators on human health,* ISBN: 90-73361-69,9, Greenpeace Research Laboratories, University of Exeter, UK

Autret, E., Berthier, F., Luszezanec, A. & Nicolas, F. (2007). Incineration of municipal and assimilated wastes in France: Assessment of latest energy and material recovery performances, *Journal of Hazardous Materials* B139, 569-574

Belgiorno, V., De Feo, G., Rocca, C. D. & Napoli, R.M.A. (2003). Energy from gasification of solid wastes, *Waste Management* 23, 1-15

Blahos, L. (2000). *Plasma Physics, the Fourth State of Matter,* Giolas Editions, 1-12

Calaminus, B. & Stahlberg, R. (1998). Continuous in-line gasification/ vitrification process for thermal waste treatment: process technology and current status of projects, Waste Management 18 (1998) 547-556

Carabin, P. & Holcroft, G. (2005). Plasma resource recovery technology converting waste to energy and valuable products, in: Proceedings of the 13th Annual North American Waste to Energy Conference, NAWTEC13, 71–79, Article number NAWTEC13-3155

Carabin, P., Palumbo, E. & Alexakis, T. (2004). Two-stage plasma gasification of waste, in: Proceedings of the 23rd International Conference on Incineration and Thermal Treatment Technologies, Phoenix, AZ, USA, May 10–14.

Circeo, L. (2007). Plasma Arc Gasification of Municipal Solid Waste, EPA Region 4 Clean and Sustainable Energy Conference Embassy Suites Hotel at Centennial Olympic Park, Atlanta, GA

Deriziotis P. (2004). Substance and perceptions of environmental impacts of dioxin emissions. M.S. thesis, Columbia University (data by U.S. EPA)

Directive 2000/76/EC of the European Parliament and of the Council of 4 December 2000 on the incineration of waste.

European Commission, (2006). *Integrated Pollution Prevention and Control Reference Document on the Best Available Techniques for Waste Incineration*

Gagnon, J. & Carabin, P. (2006). A torch to light the way: plasma gasification technology in waste treatment, Waste Management World 1, 65–68

Gidarakos E. (2006). *Hazardous Waste, Management, Treatment, Disposal,* Zigos Editions, Thessaloniki

Gomez, E., Rani, D.A., Cheeseman, C.R., Deegan, D., Wise, M. & Boccaccini, A.R. (2009). Thermal plasma technology for the treatment of wastes: A critical review, *Journal of Hazardous Materials,* 161, 2-3, 614-626

Groi, B., Eder, C., Grziwa P., Horst, J. & Kimmerle, K. (2008). Energy recovery from sewage sludge by means of fluidised bed gasification, *Waste Man*agement 28, 1819–1826

Huang, H. & Tang, L. (2007). Treatment of organic waste using thermal plasma pyrolysis technology, *Energy Conversion and Management,* 48, 1331–1337

Institution of Mechanical Engineers. (2007). Energy *from waste, A wasted opportunity?,* United Kingdom

Juniper Consultancy Services Limited. (2006). Independent Waste technology Reports, Bathurst house, Bisley GL6 7NH, England

Klein, A. (2002). Gasification: An alternative process for energy recovery and disposal of Municipal Solid Wastes. MS Thesis, Columbia University

Kuo, Y.-M., Wang, C.-T., Tsai, C.-H. & Wang, L.-C. (2009). Chemical and physical properties of plasma slags containing various amorphous volume fractions, *Journal of Hazardous Materials* 162 (1), 469-475

Leal-Quirós, E. (2004). Plasma Processing of Municipal Solid Waste, *Brazilian Journal of Physics,* 34, 4B, 1587-1593

Lemmens, B., Elslander, H., Vanderreydt, I., Peys, K., Diels, L., Oosterlinck, M. & Joos, M. (2007). Assessment of plasma gasification of high caloric waste streams, *Waste Management* 27 (11) 1562-1569

Malkow, T. (2004). Novel and innovative pyrolysis and gasification technologies for energy efficient and environmentally sound MSW disposal, *Waste Management* 24, 53-79

Moustakas, K., Fatta, F., Malamis, S., Haralambous, K.-J. & Loizidou M., (2005). Demonstration plasma gasification/vitrification system for effective hazardous waste treatment, *Journal of Hazardous Materials* B123 120-126

Moustakas, K., Xydis, G., Malamis, S., Haralambous, K.-J. & Loizidou M. (2008). Analysis of results from the operation of a pilot gasification / vitrification unit for optimizing its performance, *Journal of Hazardous Materials,* 151, 473-480

Mollah, M.Y.A., Schennach, R., Patscheider, J. Promreuk, S. & Cocke, D.L. (2000). Plasma chemistry as a tool for green chemistry, environmental analysis and waste management, *Journal of Hazardous Materials* B 79 301-320

Niessen, W. (2002). *Combustion and Incineration Processes,* Marcel Dekker Inc.

Radian International LLC (2000). *A Comparison of gasification and incineration of hazardous wastes,* DVN 99.803931.02, Austin, Texas

Rezaiyan, J. & Cheremisinoff N. (2005). *Gasification Technologies, A Primer for Engineers and Scientists,* Taylor & Francis Group, LLC

Sheng, H., Wang, R., Xu, Y., Li, Y. & Tian, J. (2008). AC plasma arc system for pyrolysis of medical waste and POPs: Paper #77 Air and Waste Management Association - 27th Annual International Conference on Thermal Treatment Technologies 2, 605-612

Yassin, L., Lettieri, P., Simons, S.J.R. & Germana A. (2009). Techno-economic performance of energy-from-waste fluidized bed combustion and gasification processes in the UK context, *Chemical Engineering Journal* 315-327

7

Effective Municipal Solid Waste Management in India

Sunil Kumar
Scientist, National Environmental Engineering Research Institute (NEERI), Council of Scientific and Industrial Research (CSIR), Kolkata Zonal Laboratory, I-8, Sector "C", East Kolkata, New Township, Kolkata, 7000 107 India

1. Introduction

Indian urban dwellers generate 0.2- 0.6 kg per person per day resulting into a national total generation of nearly 105,000 metric tons of solid wastes per day. The country's largest cities collect between 70-90% of total wastes generated, while smaller cities and towns usually collect less than 50% (Kumar, 2009). Uncollected wastes accumulate on the streets, public spaces, and vacant lots, sometimes creating illegal open dumps. Residents can also simply throw their wastes at the nearest stream or burn them. Uncollected wastes, and residents' actions to deal with them, create pollution problems and pose risks to human health and the environment.

Cities spend US $11.60 - 34.90 per metric ton in waste collection, transportation, treatment, and final disposal. Most of this cost is spent on collection (60-70 %), while transportation requires 20-30 %, and final disposal less than 5 %. New Delhi, the national capital, for instance, spends 71% in collection, 26 % in transportation, and 3 % in final disposal (Kumar, 2009). Virtually all the country's collected wastes are disposed of at open dumps, which are the cheapest option available. Despite their low cost, open dumps is a source of land, water, and air pollution, as well as public health hazards.

Waste collection methods vary from city to city, and even within each city. Door-to-door collection is not widely practiced. This collection method exists where residential associations hire private scavengers to perform it. Wastes from narrow residential and commercial lanes, and areas with high traffic are often not collected. Even though India's Supreme Court ruled that municipalities should offer door-to-door collection (the Indian Supreme Court is quite powerful and plays a slightly different role than the US Supreme Court), progress to comply with this ruling has been slow (Kumar, 2009).

Slums and squatter areas often suffer from sporadic or no waste collection at all. Many low-income individuals lack toilets, and urinate and defecate on the streets or open spaces. Open defecation and disposal of sewage and garbage from such settlements needs proper attention. A large number of cows roam the streets in Indian cities, and the dung they generate is not properly managed (Kumar 2009; http://www.waste-management-world.com/index/display/article-display.368989.articles.waste-management-world.markets-policy-finance.2009.09.waste-market-potential-in-india.html).

In most cities, waste collection is inefficient. Residents usually leave wastes in front of their homes for pick up by the sweepers. Wastes are often scattered by human scavengers searching for recyclables, as well as by cows searching for food. When garbage is scattered, it must be swept by the sweepers, picked up, and loaded onto their collection vehicles (wheelbarrows, carts, and various types of vehicles) and taken to the community waste storage sites. Each neighborhood has at least one masonry unit where residents and/or street sweepers bring the wastes for storage. Most often, street sweepers simply dump the wastes on the floor of these structures. At the structures, human scavengers salvage materials, and cows and goats look for food to eat. Even though human and animal scavenging reduces the amount of wastes that need to be transported and disposed of, these activities present health risks to the animals and to human health. The cows feeding from garbage sometimes eat plastic items, eventually killing them. And the waste picker's daily contact with garbage increases their risks of suffering injuries and illness. The residues of human and animal scavenging activities are picked up from the floor and then loaded onto the vehicles that transport the wastes to the final disposal sites. Sweeping scattered wastes and picking them from the floor twice during the collection process requires considerable effort and time by municipal collection crews, ultimately lowering their productivity.

Cities usually lack recycling programs, but a large number of waste pickers recover recyclables from wastes. It has been estimated that up to 1 million individuals make a living from scavenging activities throughout India. Scavengers recover any materials and items that can be reused and recycled: paper, plastics, metals, and so on. Several cities have composting programs, but they often process mixed wastes, which produce low-quality compost. Thus, the situation has aggravated in many cities. However, a few municipalities initiated activities to improve the situation in the light of MSW (Management and Handling) Rules, 2000

2. Effective MSW Management in India

Surat was transformed in 18 months from one of India's filthiest cities to one of its cleanest. Any strategic action plan for a city should be based and try to replicate Indian success stories.

Surat followed the following strategies:

- Developed a vision. Morale was built form the bottom up. Sweepers colonies were the first to be cleaned. It aimed to have an administration with a human face;
- The Health Officer's workplaces were cleaned;
- They started to clean the dirtiest areas;
- One task or topic at a time was tackled, and successful practices and work routines and reporting systems were put in place before starting on reform of another problem area;
- The worst problems and worst areas were decided collectively by all the senior staff and inspectors;
- Field work was a must all morning for all staff. The slogan "From AC to DC" From Air-Conditioned to Daily Chores was used;

- There were daily review meetings by the top city officer every afternoon from 3- 4 PM, with all departments present so that problems could be aired, discussed and solved on the spot;
- Both responsibility and financial authority were fully delegated to each of the zonal chiefs, who were able to take prompt decisions and solve problems immediately using their best judgment.

After a period of internal reform and only after they reached a high level of city cleaning services, Surat and Calcutta began a system of "additional cleaning charges" for residents that did not comply with the new system. These charges are higher than the former "fines" and can be collected on the spot. However, cities should not punish residents for throwing wastes on the roads if cities cannot regularly and properly clean all garbage points themselves. Firmness and fairness are also important. In Surat, when persistent defaulters such as large commercial establishments refused to pay heavy administrative charges, their shutters were downed until they did. There cannot be one rule for petty traders and another for the rich and powerful.

Learning from Others Best Practices

The Bangalore City Corporation benefited immensely from a Best Practices Workshop for Solid Waste Management, organized in May 2000 by the CM-appointed Bangalore Agenda Task Force (www.batf.org or www.blrforward.org). Nine top performers ("navaratnas") from all over India were invited to present their success stories in 9 fields, including primary collection, recycling, secondary collection and monitoring, and innovative slum clean up (www.blrforward.org).

Some of the "navaratnas" were invited to start demonstration projects in Bangalore. The city managers of Gujarat have created a forum for sharing information between themselves in order to learn from each other. Their publication on Best Practices is worth studying carefully for successful ideas in several areas.

Similarly, other cities like Pune have also initiated a lot of activities for improvement in the existing MSW management system. In the light of existing MSW (Management and handling) Rules, 2000, the Pune city has converted its open dumped site into partially sanitary landfill. Other initiatives on recycling of recyclables and improvement in the existing collection system have also been implemented.

3. Conclusions

Keeping in view of the judicial intervention, the municipalities have started a lot of activities now to improve the existing MSW management system. However, still a long way has to go to achieve sustainable waste management in India. The existing MSW rules are being modified and the Union Government has provided lot of funds in this sector and a paradigm shift is expected under 11th plan.

4. References

[1] Kumar, S., Bhattacharyya, J. K., Vaidya, A. N., Chakrabarti, T., Devotta, S, Akolkar, A.B. Assessment of the status of municipal solid waste management in metro cities, state capitals, class I cities, and class II towns in India: an insight, Waste Management 29 (2009) 883–895

[2] http://www.waste-management-world.com/index/display/article-display.368989.articles.waste-management-world.markets-policy-finance.2009.09.waste-market-potential-in-india.html

[3] www.blrforward.org

8

Thermal Conversion Technologies for Solid Wastes: A New Way to Produce Sustainable Energy

Jaime F Puna and M Teresa Santos
High Institute of Engineering of Lisbon, Chemical Engineering Department
Portugal

1. Introduction

The solid waste generation is an important environmental problem, because it grows at a rate that exceeds the ability of natural environment assimilation and the treatment capacity available. Nowadays we need to reduce the consumption of raw materials and to increase the rate of recovery and reuse of waste materials.

An essential component in many integrated solid waste management systems is thermal conversion. This kind of technology allow to obtained volume reduction and energy recovery. The energy produce by solid waste treatment contribute for the use of less fossil fuels and can help meet renewable energy targets, as a consequence of global warning problem, and contribute significantly to achieving Kyoto Protocol objectives.

As it is knowledge of the scientific community, the integrated solid waste treatment follows a hierarchic management strategy, which is sequential and obeys to some steps, in decreasing order of waste best destination (Puna, 2002).

In the nineties the waste management hierarchy usually was composed by: source reduction, recycling, waste combustion and landfilling. Nowadays waste management hierarchy is more complete because the use of chemical and biological treatments (aerobic and anaerobic). The development of a proper waste management system depends on the availability data on the characteristics of the waste stream, performance specifications for alternative technologies and cost information. (Tchobanoglous *et al.*, 1993). The United Kingston and United State of America often disregard waste incineration on future waste management systems, but other countries like Switzerland, Japan and Denmark incinerate more than 65% of municipal solid waste (Damgaard *et al.*, 2007). There are advantages and disadvantages with all treatment options.

As mentioned before, the wastes have to be submitted to one or more waste solid treatment methods and technologies. These treatment methods actually available and suitable to treat those solid wastes are classified attempting to their dangerousness (no dangerous and dangerous wastes) (Puna, 2002).

2. Fundamentals of thermal processing

Thermal processing of solid waste can be defined as the conversion of wastes into gaseous, liquid and solid production, with or without energy valorisation (Tchobanoglous *et al.*, 1993).

<u>*No dangerous wastes*</u>:	<u>*Dangerous wastes:*</u>
Physical and Chemical treatments Biological treatments: • Aerobic Digestion; • Anaerobic Digestion. Thermal treatments with energetic valorisation: • Incineration.	Physical and Chemical treatments Biological treatments Thermal methods with energetic valorisation: • Incineration; • Co-incineration in cement furnaces; • Pyrolysis by plasma with vitrification.

The mains objectives in the thermal treatment process of solid waste are the follows (Oliveira, 2005):

- Destruction of the organic components of wastes, specially the dangerous ones;
- Reducing their volume;
- Obtain solid and/or gaseous inert products;
- Achieve a significant energetic valorisation.

On the contrary of biologic, physical and chemical technologies, the destruction of dangerous contaminators by heat is much less dependent of the waste specificity. While the chemical and biological processes needs, for each kind of waste, particular operating conditions (contact time, atmosphere regulation where the reaction occurs, suitable reactants, etc.), in thermal treatment methods, its sufficient guarantee that certain temperatures are achieved in a minimum gap time, in order to consider that all initial organic molecules will be destroyed (Higgins, 1989).

For the heavy metals, the situation is more complex, because these substances, when they enter in the solid wastes, they will out in a natural way on the liquid and gaseous effluents, with the considerable risk that some of them can be volatilize during process and causing severe environmental impacts.

The thermal methods are a final solution for most of dangerous and no dangerous solid wastes, when isn't possible treat them by biological, physical and chemical techniques. However, the thermal methods are an important component in many solid waste integrated systems.

The more important thermal methods that have been used for the recovery of usable conversion products are: combustion, gasification and pyrolysis (Table 1) (Peavy *et al.*, 1985; Tchobanoglous *et al.*, 1993).

Process	Conversion product	Pre-processing
Combustion (Incineration)	Energy in the form of steam or electricity	None in mass-fired incinerator
Gasification	Low-energy gas	Separation of the organic fraction, particle size reduction, preparation of fuel cubes or other RDF
Pyrolysis	Medium-energy gas, liquid fuel, solid fuel	Separation of the organic fraction, particle size reduction, preparation of fuel cubes or other RDF

Table 1. Thermal process for the solid waste treatment (adopted of Peavy *et al.*, 1985).

The first one is an exothermic process, which means, its release spontaneously, significant energy to the process become autonomous and, also, to export energy in heat, or, most important, in electric energy. On the contrary, the gasification and pyrolysis processes are endothermic, which means, it's necessary to supply thermal energy to perform the pyrolysis reactions.

In order to judge the efficiency levels of thermal treatment techniques, normally, its employ some treatment parameters, described as follows (Russo, 2005).

DRE-Destruction and Removal Efficiency: Represents a measure of destruction and removal efficiency for a specific or a whole of specific dangerous substances, present in the solid wastes:

$$DRE = \frac{m_e - m_s}{m_e} \times 100\% \tag{1}$$

where:

m_e - contaminant mass at the incinerator inlet;

m_s - contaminant mass in the combustion gases at the incinerator outlet.

A value of 99.99% in the DRE meaning that, in maximum, only 0,0001.m_s can persist in the process combustion and flows through the combustion gases, after treatment. By another hand, a higher value of DRE doesn't meaning necessarily the elimination of a specific dangerous compound. On the contrary, it means that exists in very small concentrations on the combustion gases.

Burning index: In a waste containing several organic substances, its necessary settle which are the main dangerous organic compounds, the POHC (principal organic hazardous compound). To establish a DRE of 99.99%, its extremely important identify in each complex matrice waste, which is its POHC. With the efficient destruction of that compound, it's easy to achieve the optimal conditions to assure the elimination of the others chemical products, also dangerous. A suitable criteria for POHC determination is the burning index (I), defined as follows:

$$I = C + \left(a/H \right) \tag{2}$$

where:

C - concentration of each organic chemical substance in the waste;

a - constant with the value of 100 kcal/g;

H - combustion specific heat (kcal/g).

A higher value of index I for a specific compound indicates a great difficulty in their elimination, through incineration. So, bigger concentrations or small combustion specific heats are indicators of significant difficulties in the dangerous compounds elimination. These whole of criteria allows to estimate if will occur problems in the destruction of POHC.

DE -Destruction Efficiency: This parameter is more representative of elimination and removal efficiency process and, it's quantified by the following equation:

$$DE = \frac{m_e - m_s}{m_e} \times 100\% \tag{3}$$

where:

m_e - contaminant mass at incinerator inlet;
m_s - sum of all products masses formed in the combustion gases (ashes, combustion gases, slag's and products that remains in the washing systems and filters), produced by that contaminant.
An important aspect related with these parameters it's the possibility to know, in advance, which are the minimum temperatures and compositions atmospheres to achieve a minimum DE of 99.99% or 99.9999% in the presence of dioxins and furan's, identified as POHC.

3. Combustion treatment technologies for solid wastes

3.1 Incineration

The incineration consists in mass combustion technique of solid wastes, which are admitted to an incineration furnace. Inside the furnace occur the combustion of organic wastes including the dangerous ones with air excess to promote mixing and turbulence, in order to ensure a safety and completely burn of those substances. As co-products of this process, occurs the formation of ashes and solid slag's. The first ones are completely inertized in cement matrices for succeeding compactation in landfills and, the second ones, are valorised, separating the metals from the inerts. The metals are recycled to the recycling industry and the inerts are normally used on civil construction, such as road flooring, landfills covering, etc. (Puna, 2005). There are two incineration processes, regarded with energetic valorisation. As a great advantage of this process, it's possible to produce large quantity of electric energy, and therefore, it makes profitable this process, becoming autonomous and supplier of electricity.

It is important refer the control of main process variables of furnace incinerator, like, temperature, waste time residence on furnace, qualitative and quantitative analysis of solid wastes in admittance. Besides, the incineration and co-incineration processes are very restrictive in the admittance of several dangerous solid wastes, due to the legal limit values of gases that are produced in the incineration furnace and emitted to the atmosphere, which they are severally rigid, like, Dioxins, Furans, PCB's and Heavy Metals (Brunner, 1994).

Basically, the incineration furnace is a combustion chamber where, the solid wastes chemical elements (carbon, hydrogen and, if exists also in the wastes, sulphur) are burned to produced combustion gases, especially CO, CO_2, H_2O, NO_x and, if it is the case, SO_2. With fewer proportions, it's produced, also, acid gases like HCl and HF and, last but not least, heavy metals and macromolecules with high stability and higher molecular weight (Dioxins, Furan's and PCB's) (Freeman, 1988). The main elementary reactions of solid wastes in the combustion process at the incinerator are the follow ones:

$$C + O_2 \rightarrow CO_2 \qquad 2H_2 + O_2 \rightarrow 2H_2O \qquad S + O_2 \rightarrow SO_2$$

The dedicated incineration works like an appropriate industrial infra-structure, which uses, to operate the incineration furnace, a secondary fuel, like natural gas, propane or fuel oil to improve and maintain the combustion of solid wastes. However, the main source of fuel is the solid wastes due to its higher specific heat. It's estimated that, the solid wastes can substitute the use of secondary fuel until a percentage of utilisation about 40-50% (Formosinho *et. al.*, 2000).

The rate of organic molecules destruction depends of the high temperature inside in the furnace and the time residence of gases combustion in the incinerator. Normally, a temperature higher than 900 °C and, a time residence between 2 and 5 seconds, with an excess of air (oxygen) higher than 6% is sufficient to ensure the destruction of all organic molecules.

The figure 1 shows a *flow-sheet* of municipal solid waste dedicated incineration process. Valorsul is an integrated urban solid waste system, which include a incinerator, in order to burn urban solid wastes. This incinerator is located in Loures, near Lisbon, Portugal and, it operates since 1994 (Puna, 2002).

In the incineration of dangerous solid wastes with more than 1% of halogen organic compounds in their composition, expressed in chlorine, the temperature in the incinerator has to achieve 1100ºC and the residence time of combustion gases must be, at minimum, 2 seconds (Brunner, 1994; European Legislation, 1994). The main final products in the incineration process are, the combustion gases. The appearance of CO in the combustion gases results from the inefficient burning of solid wastes with, most probably, a less air excess. To ensure the reduction of CO, it's important to increase the air flow and, consequently, the excess of oxygen. With the combustion gases, flows particles, specially, those with a diameter smaller than 10 μm (Russo, 2005). As remaining solid wastes, the incineration process produces ashes and slags, which will be characterized more ahead.

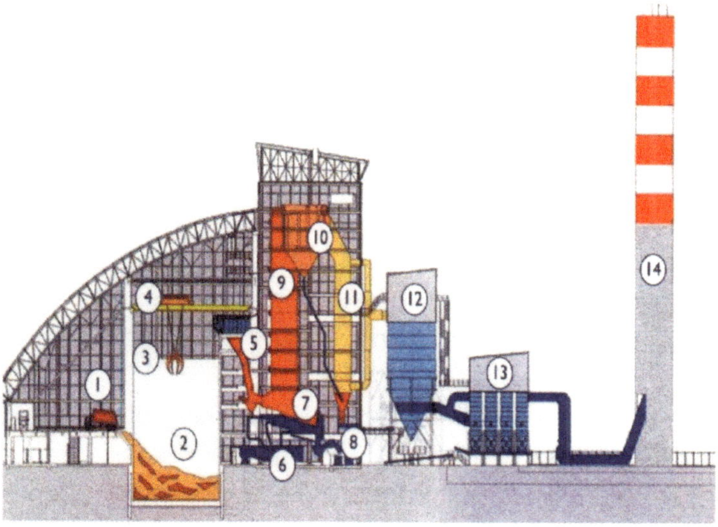

1. Discharge platform
2. Ditch
3. Claw
4. Roller bridge
5. Hopper waste feed
6. Combustion air in
7. Combustion chamber
8. Slag's extractor
9. Boiler
10. Overheater
11. Heat saving
12. Wet scrubber
13. Sleeves filters
14. Stack

Fig. 1. Diagram of Valorsul's municipal solid waste dedicated incinerator (Levy & Cabeças, 2006).

Down the incineration chambers, the gases combustion have to be submitted to physical and chemical treatments, in order to ensure contents below to their respective legal limits. The main concerns are to achieve very low limits regarded with CO and volatile organic compounds (VOC's), which are proceeding from incomplete combustion with low air excess), NO_x (produced due to high temperatures in the furnace interior), acid gases (HF, HCl e SO_2), produced due to the presence of halogen (F, Cl) and Sulphur atoms in the solid wastes composition, heavy metals (Cu, Cr, Cd, Be, Mn, Hg and As), Dioxins/Furan's/PCB's, and particles.

Today, the European Directive on waste incineration (76/2000/EC), overcome to the Portuguese legislation, by the DL n.º 85/2005) considers a limit value for Cr together with

other eight heavy metals (Sb + As + Pb + Cr + Co + Cu + Mn + Ni + V) equal of 0.5 mg/Nm^3 without taking into account the different toxicity of total Cr compared to Cr^{VI} (Cocarta, *et al.*, 2007).
Table 2 shows monitorization data related with treated gases combustion in the stack outlet of Valorsul's incinerator (CTRSU - S. João da Talha, Loures, Portugal), as well, the Portuguese legislation (Port. 286/93 and DL 85/2005), which establishes legal limits for each pollutant released to the atmosphere.
An incinerator powerplant must be gifted also, with sophisticated systems for treating the combustion gases; a turbine to convert thermal energy proceeded from super heated steam water into mechanical energy and, also, an alternator to convert this mechanical energy into electric energy. This last equipment is extremely important to profit the heat release from the combustion (exothermic process) to produce super heated steam water for subsequent production of electricity.

Atmospheric pollutant	Port. 286/93	CTRSU	DL 85/2005
HF (mg/Nm^3)	2,0	1 / 0,8	1
HCl (mg/Nm^3)	50	20 / 20	10
SO_2 (mg/Nm^3)	300	50 / 50	50
Particles (mg/Nm^3)	30	20 / 10	10
NO_x (mg/Nm^3)	n.a.	250	200
CO (mg/Nm^3)	100	50 / 50	50
COV (mg/Nm^3)	20	10 / 10	10
Dioxins/Furan's (ng/Nm^3)	n.a.	0,1/ 0,05	0,1

Table 2. Monitorization of atmospheric pollutants at stack outlet, in the Valorsul's incinerator (CTRSU), as well, their respective legal limits.

The co-products resulted from waste solid combustion in incineration chambers are easy to compact and to store, due to their reduced volume and, they are classified as ashes and slags. Both are considered as remaining solid wastes. The ashes are submitted to an inertization process before their deposition in controlled landfill, while the slags are submitted to a valorisation process, which consists in separate the metals from the inerts. The metals can be conducted to the recycling industry and, the inerts can be used as covering of landfills, flooring roads, etc. The typical composition of slag's are 40% of SiO_2, 10% to 20% of Al_2O_3 and Fe_2O_3, 15% of H_2O and oxides, phosphates and sulphates, with a content below to 6%. By another hand, the ashes typical composition is 15% of sulphates, 13% of chlorides, 7-8% of SiO_2, 4-7% of Al and other alkaline and heavy metals with a proportion not higher than 5% (Puna, 2005).
There is still the production of volatile ashes, proceeding from the treatment of combustion gases in the depuration process. These ashes have to be also, submitted to an inertization procedure, before their deposition in landfill. Table 3 shows the main processes data of Valorsul's dedicated incinerator of urban solid wastes.

Furnace temperature	900 °C - 1200 °C
Wastes nominal calorific power	7600 - 7900 kJ/kg
Waste reception	662 000 ton./year
Solid slag's production	200 kg/ton. waste
Ashes production	30 kg/ton. waste
Electric energy production	587 kWh/ton. waste for 150.000 habitants
Water steam	Over-heated (T>300°C)
Turbine strength	Condensation, 50 MW

Table 3. Processes data of Valorsul's dedicated incineration (S. João da Talha, Loures, Portugal).

In summary, it's possible to express the main processes characteristics which define the dedicated incineration of solid wastes:

- The temperature in the inside of furnaces has to be controlled between 900°C and 1200°C;
- The combustion gases must have a short time residence in the chamber incineration, at high temperature, to avoid the considerable production of Dioxins/Furan's/PCB's. This can be achieved with 2-5 seconds for a temperature of 900°C and with an oxygen excess higher than 6% (Brunner, 1994);
- The admittance of solid wastes in the incineration chambers will only be possible with chlorine contents below 1% (w/w), to minimize and avoid, once again, the production of Dioxins/Furan's, which occurs in the combustion process, at high temperatures (below 900°C) and with high contents of Chlorine (Puna, 2002);
- A incineration powerplant has to be a installation to treat the combustion gases and, has to treat also, the co-products of incineration process (ashes, slag's and volatile ashes resulted from combustion gases treatment process);

Fig. 3. Inside of an incineration chamber in Valorsul's incinerator (S. João da Talha, Loures, Portugal).

- The dedicated incineration will only be an energetic valorisation process if the energy spontaneously released in the combustion chambers could be profited to produce superheated steam water. The high enthalpy of steam water at high temperature has a significant economic value and, it can be used to produce electric energy, in order to supply electricity, not only for the own incineration powerplant, but, more important, supply to the national electric network;
- The main incomes of an incineration unit are the admittance of solid wastes and, the supply of electricity to the electric network. Besides that, it's possible to avoid costs related with the purchase of electric energy coming from the operator's supplier electricity.

It's possible to see in figure 4, a typical diagram of industrial solid waste dedicated incinerator with energetic valorisation. The figure 3 shows the inside of an incineration chamber.

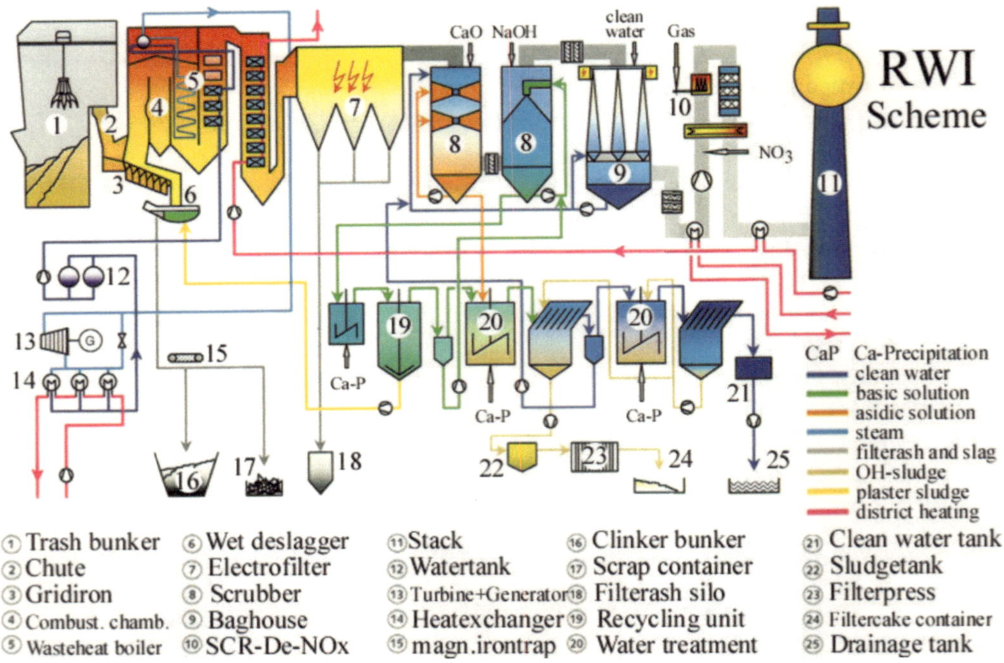

Fig. 4. Typical diagram of an industrial solid waste dedicated incineration process with energetic valorisation (http://www.sbg.ac.at/ipk/avstudio/pierofun/waste/residual/step-1.jpg).

3.2 Co-incineration

Co-incineration of solid wastes is, also, a mass combustion process, which occurs, mostly, in cement furnaces, simultaneously with the clínquer production. The difference results that this is a "dry" process, because there isn't, in any part of the process, the flowing of water/steam water as industrial utility (Puna, 2002). By another hand, the co-products (dangerous ashes, slag's and several heavy metals) are all incorporated in the hardness cement crystalline structure, becoming inerts. Regard the figure 5, which shows a typical scheme of a co-incineration furnace of dangerous solid wastes.

The co-incineration is a thermal destruction technique applied to several industrial solid wastes and, these wastes are valorised as raw-materials and/or fuels. Several industrial process which works at high temperatures, can be used for waste elimination, taking advantage of their calorific power. The elimination of industrial wastes with a calorific power minimum of 5000 kJ/kg in an industrial process can be considered as a technique with energetic valorisation (Oliveira, 2005). This method of thermal destruction of solid wastes has several benefits, even as an environmental perspective, but also, and more important, as an energetic perspective, because enables to substitute fossil fuels, like fuel oil, natural gas, etc., by dangerous wastes. These wastes are valorised in two forms: can be used as raw materials and, as fuel supplier to burning furnaces (Puna, 2002). However, the substitution of fossil fuels by dangerous wastes cannot increase the atmospheric pollutants emissions, resulting from combustion processes, in comparison with the normal use of conventional fuels. (Pio *et. al.*, 2003).

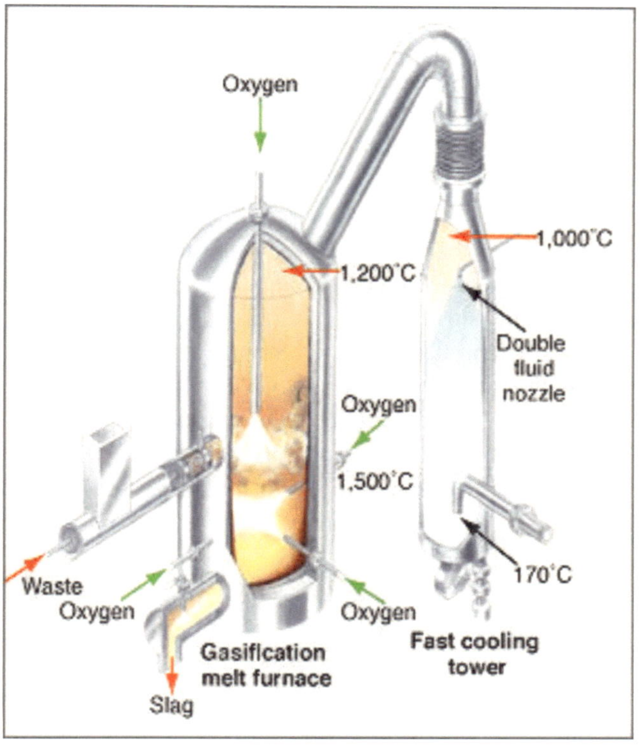

Fig. 5. Typical diagram of a co-incineration process of dangerous solid wastes (http://www.sumitomo.gr.jp/english/discoveries/special/images/100_07.jpg).

The co-incineration can be conducted in different industrial processes, like, for instance, in cement furnaces, or in industrial boilers. Nevertheless, the co-incineration in cement furnaces is considered the most efficient process of co-incineration, especially to dangerous solid wastes (Formosinho *et. al.*, 2000). The combustion of wastes in cement furnaces occurs at the same time with the production of clínquer (cement product intermediate). The main characteristics in the co-incineration process of solid wastes in cement furnaces are the follow ones (Scoreco, 1997):

- Thermal valorisation method, alternative to dedicated incineration, only applied to burning wastes with higher calorific power, like, used oils and fatty slush's of industrial wastewaters treatment units;
- It's necessary a previous treatment for wastes before entering in the cement furnace, through physical and chemical process (impregnation, melting, centrifugation, fluidization);
- The industrial solid wastes are burned as fuel, with oxygen from air, in a mass combustion process, with temperatures between 1400ºC-1500ºC;
- The combustion gases achieve maximum temperatures near from 2000ºC in the main burner and stay at temperatures higher than 1200ºC in the second burner, with time residences between 4-8 seconds;
- The wastes admitted to co-incineration in cement furnaces cannot be contain chlorine contents higher than 1% (w/w), due to the significant production of dioxins/furan's, when the combustion gases are cooled faster at the outlet of clínquer furnace;
- These operating conditions are crucial to reduce and avoid the production of those macromolecules with higher molecular weight. It's important to remind that, dioxins/furan's molecules are produced, in a combustion process, with temperatures between 250-900ºC and, with significant contents of Chlorine in the solid wastes. Besides that, the chlorine is harmful to the consolidation of cement structure, raising several weakness;
- It's necessary also, like in the dedicated incineration, the treatment of combustion gases, before they go out to the atmosphere, with temperatures between 150-200ºC. To achieve this purpose, the process of cooling the combustion gases has to be very fast and, in a temperature gradient between (1000-1200)ºC until (150-200)ºC;
- The temperatures profile and the time residences are higher than any other combustion process, like dedicated incineration;
- Basically, a cement furnace is a place with optimal conditions to burn and eliminate any organic waste with capacity to be submitted for incineration;
- It's extremely important the control of operating parameters, like, temperature, oxygen content and time residence of combustion gases, to ensure an efficient and safety burning of solid wastes, mainly, the dangerous ones;
- The thermal energy to feed the furnace is obtain by a variety of auxiliary fuels, but with large preference to coal and/or pet-cock, with very low contents of sulphur, to avoid the production of SO_2;
- The burning dangerous industrial wastes with high calorific power can replace the coal, as fuel to supply the co-incineration furnace, until 40%, with, mainly, used oils, solvents and organic slush's.

Figure 6 shows some pictures of a cement furnace.

It is also important refer that, the combustion gases, before they go out to the atmosphere, are previously treated by physical and chemical appropriated process, like in dedicated incineration, in order to maintain the air quality and, therefore, assure that the gaseous emissions could be above their legal emission limit values, defined in the European and Portuguese legislation. Heavy metals, Dioxins, Furans and PCB's are treated by adsorption with activated carbon, while the acid gases (HCl, HF and SO_2) are treated by chemical reaction with lime milk ($Ca(OH)_2$) (Russo, 2005).

Fig. 6. Picture of inside (left) and outside (right) of a co-incineration unit in cement furnace used to burn dangerous industrial solid wastes (Puna, 2002).

The NO_x gases are treated by injection, without catalyst, of ammonia aqueous solution, producing N_2, while the almost particles are filtered with sleeves filters of higher efficiency or, with electro filters. The wastes generated by these processes are called flying ashes and they are covered in safety and appropriate landfills, after an inertization process.
This group of combustion gases treatment is performed at clínquer furnace exit and, they are the same methods, with the same technologies used in dedicated incineration, described in table 4. Table 5 identifies the emission limit values of several gaseous pollutants to the atmosphere, according with current Portuguese legislation (DL n.° 85/2005), overcome by EU legal framework.

Gaseous Pollutant	**Gas Combustion Treatment Process**
NO_x	Selective removal with injection of ammonia aqueous solution, producing N_2.
HCl, HF, SO_2	Injection of ($Ca(OH)_2$), in the gases purifying, through a *"scrubber"* process (gases washing).
Dioxins/Furan's/PCB's	Injection of activated carbon in the gases purifying, to adsorb these substances; efficient control of furnace temperature and time residence of gases combustion, with very fast cooling.
Heavy metals (As, Cd, Be, Pb, Hg, Zn, Cu, Cr)	Injection of activated carbon in the gases purifying, to adsorb these substances.
Particles/dusts	Use of sleeves filters of higher efficiency or, with electro filters.
CO/VOC's	Ensure a complete burning, supplying air in considerable excess.

Table 4. Techniques and technologies used in the combustion gases treatment (Adapted from Pio *et. al.*, 2003 and Formosinho *et. al.*, 2000).

It's interesting to perform a comparison between dedicated incineration and co-incineration thermal methods for solid wastes. Stronger and weaker aspects can be confronted in table 6. Table 7 identifies the number of industrial units in Europe, where is possible to burn dangerous industrial wastes, by co-incineration in cement furnaces.

Pollutant	Unit	Limit-Value	Pollutant	Unit	Limit-Value
Particles	mg/Nm^3	10	CO	mg/Nm^3	50
SO_2	mg/Nm^3	50	VOC's	mg/Nm^3	10
HCl	mg/Nm^3	10	Dioxins/Furan's	ng/Nm^3	0,1
HF	mg/Nm^3	1	Pb + Cr + Cu + Mn	mg/Nm^3	5
NO_x	mg/Nm^3	200	Cd + Hg	mg/Nm^3	0,2

Table 5. Limit values of gaseous pollutants emissions at incinerator stack exit (DL 85/2005).

Dedicated Incineration	Co-Incineration
Less efficiency of organic molecules destruction (1100ºC/2-5 s).	Higher efficiency of organic molecules destruction (1450ºC/4-8 s).
Can accept more contaminated wastes, like, organometallics and organochlorides.	Can't burn any dangerous industrial solid wastes with contents of chlorine higher than 1% (w/w).
It produces new remaining wastes, like ashes, slag's and washing effluents, which have to be treated also.	The ashes, slags, heavy metals and other pollutants can be fixed in the final matrice of cement, without lixiviation.
Higher energetic and economic yields, due to the electricity production and supplying.	Higher energetic yield, due to the substitution of fossil fuels, by solid wastes.
Doesn't need of any previous treatment for solid wastes, instead of co-incineration.	More restrictive related to the presence of some heavy metals in the solid wastes. Its need a previous treatment to admit the solid wastes.

Table 6. Comparison between dedicated incineration and co-incineration in cement furnaces (Adapted from Formosinho *et. al.*, 2000).

4. Pyrolysis and gasification of solid wastes

In the pyrolysis technologies, the most efficient is the PPV process, which means Pyrolysis by Plasma with Vitrification. Pyrolysis is a technology dedicated of waste destruction, which works at high temperatures, more than the typical temperatures in incineration chambers, with low oxygen, in order to avoid the combustion phenomena (Camacho, 2005). To guarantee the absence of oxygen, the wastes are decomposed in an inert gaseous atmosphere, through the utilisation of Nitrogen (N_2) (Puna, 2002). The process of pyrolysis can be defined, generally, as the chemical decomposition of organic matter by heat, in the absence of air, unlike the incineration methods. The pyrolysis processes is endothermic, on the contrary of dedicated incineration or co-incineration, because it's need to supply heat to the pyrolysis reactor in order to occur the pyrolysis reactions. If any gas is heated at higher temperatures, there are significant changes in their properties. In the range of temperatures between 2000ºC and 3000ºC, the gas molecules decompose in ionized atoms by loosing electrons. This ionized gas is called plasma (Lapa & Oliveira, 2002).

Country	Total units (B)	Units that perform co-incineration of dangerous wastes (A)	A/B (%)
Germany	46	16	35
Austria	9	7	78
Belgium	5	5	100
Denmark	1	0	0
Spain	39	6	15
Finland	3	0	0
France	47	19	40
Greece	8	0	0
Irland	4	0	0
Italy	62	5	8
Luxembourg	1	1	100
Netherlands	1	1	100
Portugal	6	0	0
U. Kingdom	15	2	13
Sweden	3	3	100
Switzerland	11	7	64
Total	**261**	**72**	**28**

Table 7. Industrial units in Europe where is possible perform co-incineration of dangerous solid wastes, in cement furnaces (Adapted from Formosinho *et. al.*, 2000).

Normally, the wastes are injected directly in the plasma, producing pyrolysis gas (essentially H_2, CO, N_2, CO_2, CH_4), and this gas can be burned in a combustion process, by incineration, in order to make profitable the entire process and to valorise it as a gas fuel, since CO and CH_4 are organic gases with high calorific power. Nevertheless, it's necessary a higher and significant annual flow admittance solid wastes to maintain the optimum operating conditions of PPV reactor and, also, to profit the all PPV system, since the production of plasma is a great consumer of thermal energy (Camacho, 2005).

The co-products of this process, specially ashes and heavy metals, are encapsulated in a vitrified matrice, to avoid the production of leachates. This vitrified matrice transform the PPV co-products in inerts remaining wastes, without any chance of occur lixiviation. This is a great advantage in an environment and public health perspectives. This vitrified matrice is called "obsidiana" and, results from the cooling of glass file-dust, which is introduced in the pyrolysis reactor, on the temperature range of 2000ºC-3000ºC (Oliveira, 2000). The glass at these temperatures is liquid and, in the cooling step, is submitted to a solidification process, covering the remaining wastes, heavy metals and other dangerous gaseous/solid substances produced in the pyrolysis reactor. These vitrified ashes have large applicability in the road flooring, landfills covering and, as additive to the cement in civil construction.

In this process, the application range of dangerous solid wastes is almost total and much more all-inclusive that the admittance wastes in incineration methods. In all thermal processes, this is the one that is considered, in an environmental point of view, the most sustainable, although the higher energetic and economic costs (Puna, 2002). The general equation of a pyrolysis process can be traduced in the following way:

$$\text{Organic matter} + \text{Heat} \rightarrow \text{Gases} + \text{Refractory metals}$$

The plasma is a special form of gaseous material, capable to conduct electricity and, it's knower as the "fourth state of matter" (solid, liquid, gas and plasma). In the state of plasma, the gas can achieve temperatures extremely high, which can change from 5 000 to 50 000 ºC, depending of its production conditions (Oliveira, 2000). In figure 7, it's possible to see a plasma jet.

Fig. 7. Plasma jet (http://paginas.fe.up.pt/~jotace/gtresiduos/plasmapirolise.htm).

This plasma is generated by the formation of an electric arch, through the cross of electric current between the cathode and the anode. Between them, a gas is injected and ionized. This ionized gas is, subsequently injected over the solid wastes. The plasma jet is produced and controlled in a torch capable to convert electric energy in heat, at higher temperature through the gas flow. In the torch, any gas rapidly reaches the plasma state. Figure 8 shows, in detail, the plasma jet production.

Basically, there are two kinds of solid waste treatment by plasma: the direct heating system and heating system with gasification chamber.

Direct heating system:

Through the plasma torch, occurs the production of an electric field of radiant energy with higher intensity, capable to dissociate the existing intra molecular bindings of solid, liquid and gaseous wastes, dangerous or inerts, organics or inorganics. So, when the wastes are submitted to the plasma jet, they loose their original chemical composition to convert in more simple compounds. Figure 9 shows a direct heated system diagram used in PPV system to treat municipal and hazardous solid wastes.

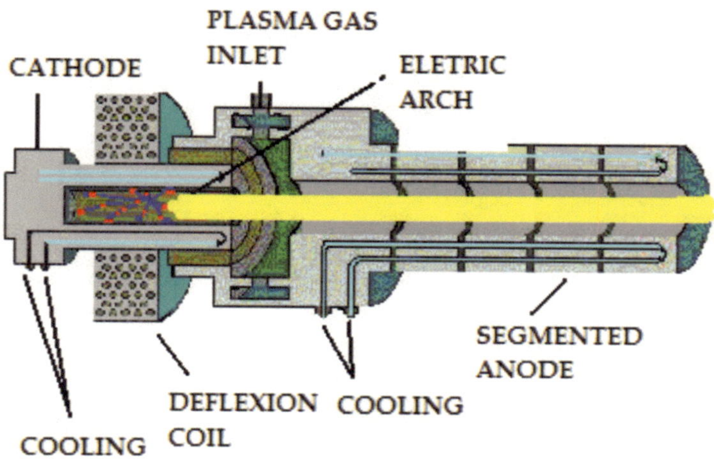

Fig. 8. Scheme of the plasma torch inside, showing the creation process of plasma jet (Adapted from http://paginas.fe.up.pt/~jotace/gtresiduos/plasmapirolise.htm).

Heating system with gasification chamber:

This system consists in two different stages of treatment. The solid wastes are injected in a first conventional gasification chamber, in order to gasify the organic compounds in a gas partially oxidized and, also, to melt the inorganic compounds. In this chamber, it's produced a gas and a liquid, which they are, subsequently decomposed in a second chamber, with a PPV reactor.

After the dissociation of all molecules, the matter is recovered in the following forms (Puna, 2002):

- **Plasma synthesized gas**, which is conducted to a combustion chamber, in order to valorise its calorific power and, to reuse the release heated, supplying into the PPV reactor;
- **Inorganic materials and vitrified silicates**, which will swim on the surface of liquid phase. These inorganic compounds, in the case of directed heating technology, were submitted to temperatures substantially higher than in the gasification chamber method.
- **Obsidiana**, which is a solid structure of higher hardness and, generally, with black colour, similar to a mineral of volcanic source. This solid contains the PPV ashes, the heavy metals and other dangerous inorganic atoms, all vitrified, without any chance of occur lixiviation. Figure 10 shows the typical aspect of Obsidiana.

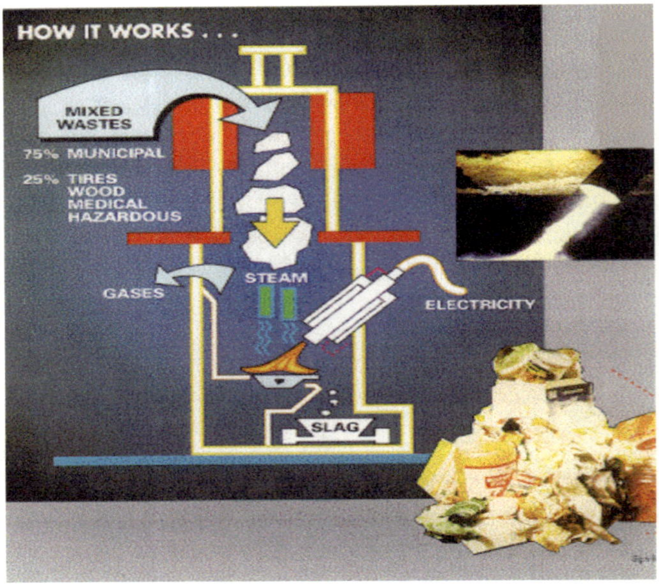

Fig. 9. Example of plasma utilisation directly over the solid wastes in the direct heating system (http://paginas.fe.up.pt/~jotace/gtresiduos/plasmapirolise.htm).

Fig. 10. Vitrified contaminants aspect after PPV reaction (http://paginas.fe.up.pt/~jotace/gtresiduos/plasmapirolise.htm).

Like other treatment techniques of industrial waste treatment, the use of pyrolysis with plasma presents advantages and disadvantages or inconvenients, as follows:

Advantages:

- PPV is a process more environmental friendly and safety, with "zero" pollutants emission or, with magnitudes lowers than those established in the environmental legal Framework related with air quality;
- Higher temperatures causes rapid and complete pyrolysis of organic wastes, melting and vitrifying certain inorganic compounds, in a high hardness structure, without lixiviation, called obsidiana;

- The plasma synthesized gas, with high calorific power, can be used in other process or, it can be submitted to combustion in order to valorise it;
- In PPV reactor, there isn't combustion of solid wastes, so, it doesn't occur the production of toxic compounds, like dioxins, furan's and PCB's;
- The gas volume obtained is substantially less then the gas volume achieved in other treatment process, like incineration, so, it's easier to be treated. The reduction rate volume from waste to gas, can be higher than 99%;
- The high temperature of PPV reactor to the molecules dissociation is produced from electricity, which is a clean energetic source;
- Enables the co-generation of energy, with production of electricity, steam and/or cold (freeze water/air conditioning).

Disadvantages:

- PPV it's a dedicated technology, requiring a high investment, due to the fact that, it can only be profitable when coupled with a thermoelectric powerplant, to supply the sufficient electricity for plasma production. It's also necessary, a significant higher and stable flow of solid wastes, which compromises any waste reduction/reutilisation/recycling policy strategy in medium/long time;
- The PPV system can't dispense a sophisticated washing gases system, as in any incineration process, especially for the retention of VOC's and acid gases, after the combustion of plasma synthesized gas;
- For different waste treatment, in particular, those containing organic matter in significant amounts, the pyrolysis techniques can't achieve great industrial development. The wastes are decomposed by pyrolysis but, after that, they are eliminated by combustion, through the incineration of plasma gas;
- The production of dioxins/furan's/PCB's in the incineration chamber, after PPV reactor, are strongly dependent of thermal recovery technologies used down the stream. It's not clearly that it can assure a significant advantage over more advanced incineration Technologies or over gasification simple techniques.

Synthesizing, the main process characteristics of PPV, are:

- It's necessary a thermal source with high enthalpy and reduced mass, which is the plasma (boiled gas at high temperatures);
- The pyrolysis temperature, the applied heating rate and the waste composition will determine the gas pyrolysis composition;
- The plasma corresponds to the fourth state of matter, the ionized gas, under temperatures between 2000ºC and 3000ºC and, it's produced by an electric discharge between the cathode and the anode, where flows a inert gas, which is injected over the wastes;
- Any organic compound, including wastes, is convertible in gas pyrolysis and in a mixture of refractable glass with PPV ashes and heavy metals, under a hardness solid structure without any material percolation;
- In this process, there isn't a final liquid phase and the higher temperatures leads to the elimination of macromolecules traditionally produced in the combustion process (dioxins, furan's, PCB's);
- The plasma is controlled under a torch, converting electric energy into heat, through the supplying of a higher amount of electricity, proceeding from a own powerplant electricity production;

- In this process, occurs the following elementary reactions in the PPV reactor, with the important auxiliary of heat generated by the plasma:

$$C + H_2O \rightarrow CO + H_2 \qquad CO + H_2O \rightarrow CO_2 + H_2$$

$$C + CO_2 \rightarrow 2CO \qquad C + 2H_2 \rightarrow CH_4$$

The average volumetric composition of plasma gas is, normally, 41% of H_2,, 30% of CO, 17% of N_2, 8% of CO_2, 3% of CH_4 and, O_2, C_2H_2, C_2H_4 with contents lowers than 0,5%. The calorific recovery in PPV process around 10.51 MJ/Nm^3 and the energetic yield is near from 612 kWh/ton of treated waste (Oliveira, 2000). Table 8 performs a comparison of the main characteristics between PPV and incineration processes. Table 9 presents the several PPV units located all over the world, especially applied in elimination toxic wastes, hazardous and radioactives.

Incineration	PPV
Combustion of wastes with air excess.	Thermal decomposition of wastes with absence of air, in an inert atmosphere, at closed reactor, with substantially higher temperatures.
System treatment conditioned to some kind of solid wastes, due to atmospheric emissions released.	System treatment applied to any kind of solid wastes.
Air volume very high.	Air volume 20 to 50 times below at incineration.
Production of ashes and slag's, which have to be treated.	Ashes and heavy metals are vitrified in a hardness solid structure, without lixiviation.
Production of several dangerous organic compounds with high stability, like dioxins, furan's and PCB's.	Destruction of organic compounds almost complete, leading to the release of pyrolysis gas, with H_2, N_2, CO, CO_2, H_2O, CH_4 and other hydrocarbons in track amounts.
In the dedicated incineration, the gases can be valorised in the production of electric energy.	The pyrolysis gas can be valorised in energy production or used in the steam production, convertible in electric energy. However, due to the gap temperatures used, the energy production can represent 20%-80% plus than the energetic consumption.

Table 8. Comparative board with the main differences between PPV and incineration processes (dedicated and co-incineration in cement furnaces) (Puna, 2002).

Treatment of different wastes	
Love Canal, Niagara Falls (EUA)	Toxic slushs elimination
BNL/EPA/COE (EUA, GB)	Descontamination of special materials
IHI (Japan)	Incinerator slag's elimination
Westinghouse/PSI (EUA)	Treatment of contaminated landfills
US Naval Surface Warfare Center (EUA)	Treatment of special wastes
New York City Harbour	Toxic slushs elimination
Elimination of toxic wastes	
DOE/USN (EUA)	Elimination of special toxic wastes
DOE (EUA)	
Bordéus city council (France)	
Kawasaki Steel Company - TEPCO (Japan)	
Ebara - Infilco (Japan) - 1993 e 1994	
Wastes of lower radioactivity	
Westinghouse Hanford (EUA)	Inertization
Sandia National Laboratory (EUA)	
DOE (OTD), Ukiah (EUA)	
British Nuclear Fuels (GB)	
DOE/Argon National Laboratory (EUA)	
Urban solid wastes and similar	
Valencia (Spain) (start up at 2002/12/10)	500 ton./day of urban wastes + tanning wastes (with Chromium recovery)
Vicenza (Italy) (start up at 2002/12/10)	500 ton./day of urban wastes + Venice channels slush's
Kuala Lumpur I (Malaysia) (start up at 2003/07/04)	360 ton./day of urban wastes + hazardous wastes

Table 9. List of industrial units with PPV waste treatment (Oliveira, 2000).

5. Environmental and health impacts control systems

In any waste solid treatment technology, there are several environmental impacts associated to the function of those units. Those impacts must be identified and, if possible, quantified, in order to know, with more precision, what are the consequences and benefits for the environmental and public health, related with the labour of those treatments. It gives, also, information to optimise all operating conditions of those systems and, to ensure the accomplishment of legal framework, especially those who are related with air quality, ecotoxicological risks for human health, quality and discharge of wastewaters, noise, remaining solid wastes, etc (Adapted from Partidário & Jesus, 1994).

The final purpose of any environmental impacts characterization is supply a sustainable development to all communities in the neighbourhoods. After all, they are the bigger beneficiaries with the implementation of any waste solid treatment systems. Particularly, for thermal methods of solid wastes, the following aspects have to be considered, in order to ensure an efficient and safety function of those units, through a continuous monitorization and control processes (Adapted from Vanclay & Bronstein, 1995 and from Wathern, 2000):

- Production of wastewaters;
- Air quality (1);
- Noise from the electromechanic equipments used;
- Deep smells resultants, essentially, from the discharge and storage of solid wastes;
- Production of remaining wastes from combustion and pyrolysis processes, respectively, ashes and slag's, and, vitrified ashes, which they have, all of them, submitted to a suitable destination for valorisation and/or treatment;
- Water quality and silts in the hydric resources involving the treatment units, as well, the respective ecosystems (2);
- Public health monitorization and analysis;
- Possibility of occur serious industrial accidents.

Notes: (1) - By law, it's obligatory to perform a continuous monitorization of gases combustion at the way out of any incinerator chimney. It must be performed also, a continuous monitorization of air quality in the neighbourhood zones, especially, in the housing ones. The parameters to be analysed are, according with respective legislation, the follow ones: SO_2, NO_x, CO, O_3, particles, HF, HCl, heavy metals, VOC's, dioxins and furan's.
(2) - Should be made a monitorization of superficial and underground hydric resources, with a schedule and a programme well constructed. The parameters to be analysed are the follow ones, according also, with the respective legislation: pH, temperature, salinity, conductivity, CQO, CBO, PCB's, PAH's, heavy metals, dioxins and furans.

However, the most important environmental parameters to be analysed is the atmospheric pollutants emissions, even in incineration or in pyrolysis processes. Table 10 shows the generic composition of gaseous pollutants emissions of PPV process and makes a comparison with the outlet gases of incineration processes. The PPV process presents less quantities of dangerous substances in their vitrified ashes, like some heavy metals (Hg, As, Pb, Zn, Cu), Nitrites, Phenol and Fluorides, unlike the incineration bottom ashes, which imply that, the PPV process is more environmental friendly. These ashes don't offer significant problems in terms of ecotoxicological risks. Additionally, it is possible to say that the provisional risks of vitrified residue material are nor significant (Oliveira, 2000).

Pollutants emissions	PPV (11% O_2)	Incineration
PCDD (ng/Nm^3)	N.D.	19 - 298
PCDF (ng/Nm^3)	3 - 10	44 - 306
PCB's (ng/Nm^3)	N.D.	2 - 7
Cd (mg/Nm^3)	0,004 - 0,03	0,6 - 0,9
Cr (mg/Nm^3)	0,02 - 0,08	0,03 - 0,1
Pb (mg/Nm^3)	0,2 - 0,6	8,4 - 15
Hg (mg/Nm^3)	N.D.	0,5 - 0,9
HCl (mg/Nm^3)	0,1 - 0,6	
HF (mg/Nm^3)	0,07 - 0,7	
NO_x (ppm)	158 - 305	169 - 246
SO_2 (ppm)	66 - 69	128 - 225
Particles (mg/Nm^3)	2,4 - 9,9	167 - 247

Table 10. Pollutants emissions from PPV and incineration processes (Oliveira, 2000).

The most dangerous pollutants for the environment and public health are the heavy metals and the dioxins/furans/PCB. Theoretically, there are 75 possible isomerous for all groups of dioxins, but only 7 develop serious toxic effects. In the same way, there are 135 possible isomerous for furans, but only 10 develop potential toxic effects. Related to PCB's, there are 209 possible isomerous, but only 11 with toxic properties (Odum, 2001).

6. The profit of energy recovered in Portuguese incineration units for solid wastes: A brief case-study.

The authors performed an economic study to achieve what were the best waste treatment methods with energetic valorisation, capable to produce large quantities of electric energy. The summary results of this study are presented here, in this book. In fact, it was possible to conclude that, attending to the description of the several techniques for waste solid treatment (dedicated incineration, anaerobic digestion and energy production in landfills), these treatment methods/technologies contributes with significant amounts of electric energy (Puna & Baptista, 2008). Table 11 shows the annual treatment capacity of solid waste for each method and, also, the respective amount of energy produced, express in MWh/year. These amounts are sufficient, by one hand, to become autonomous these waste treatment infrastructures and, by another hand, enables to export significant quantities of electricity to the National Electric Network.

Urban solid waste treatment methods	Ton./year	MWh/year
INCINERATION (a)	662000	388594
ANAEROBIC DIGESTION (b)	40000	12000
BIOGAS PRODUCTION IN LANDFILLS (c)	146000	14880
(a) – Valorsul's urban solid waste incinerator in Loures, Lisbon, Portugal (b) – Valorsul's organic valorisation plant in Amadora, Lisbon, Portugal (c) – Amarsul's landfill in Seixal, Setúbal, Portugal		

Table 11. Annual solid wastes treatment capacities and respective energy amounts produced.

This study was made, through the analysis of two waste solid treatment integrated management systems, Valorsul's, which operates in the Metropolitan Area of Lisbon, Portugal and, Amarsul's, operating in the south border of Lisbon. Both systems consist in an expressive case-study (Puna & Baptista, 2008). The first integrated system includes a dedicated incinerator and an anaerobic digestion unit. The second integrated system operates with a controlled landfill, taking advantage of the high calorific power of biogas, produced by the decomposition of the settled organic solid wastes.

Through the knowledge of these processes, it was possible to determine the nominal electric energy production for each of these treatment methods. For instance, it was important to know what is the ratio between each ton of waste treated by incineration and a kWh of

electric energy produced. In this example, 1 ton of urban solid waste treated by incineration, normally leads to 587 kWh of electric energy produced (Valorsul). These results are presented in table 12.
It's obviously that, not all the electric energy produced is exported, because some percentage is necessary to cover the energetic necessities of those waste solid treatment infrastructures. This is a great advantage because it gives them an energetic independence from the exterior. The distribution of the percentages of electric energy produced for each waste solid treatment method mentioned above depends of the respective efficiency (Puna & Baptista, 2008) and, their values are described in table 13. With this data, it's possible to estimate the annual production of electric energy exported to the National Electric Network and their respective profits or revenues, for each solid waste treatment method, attending that the kWh price to sell is 0,07€.

Treatment method	kWh/ton.waste	Eficiency (%)
INCINERATION	587	---
ANAEROBIC DIGESTION	300	30
BIOGAS PRODUCTION IN LANDFILLS	102	40

Table 12. Nominal production of electric energy.

Solid waste treatment method	Efficiency (%)	% of electric energy for own consumption	% of electric energy for exportation
INCINERATION	---	15 - 20	80 - 85
ANAEROBIC DIGESTION	30	30 - 35	65 - 70
BIOGAS PRODUCTION IN LANDFILLS	40	5 - 10	90 - 95

Table 13. Distribution of nominal production electric energy.

This price is called "green rate" because the energy produced is obtained from waste treatment, which corresponds to an energy production from renewable sources. The results obtained are presented in figures 11 and 12 (Puna & Baptista, 2008).

7. Conclusions

Now, it's important to stipulate some final conclusions, which results of all it was written here, in this chapter, highlighting the following ones:

- The incineration has a important rule in the integrated solid wastes management systems;
- If a specific solid waste could not be submitted to any other treatment process, like, biological, physical and chemical methods, then, the thermal techniques are the most suitable treatment solutions, according to the methods and technologies actually available;

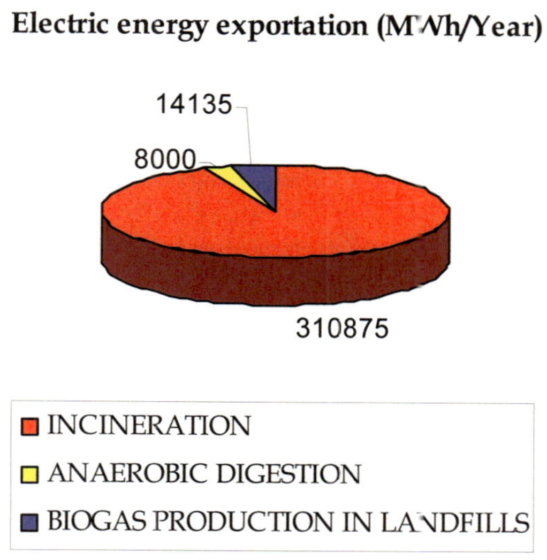

Fig. 11. Electric energy exportation.

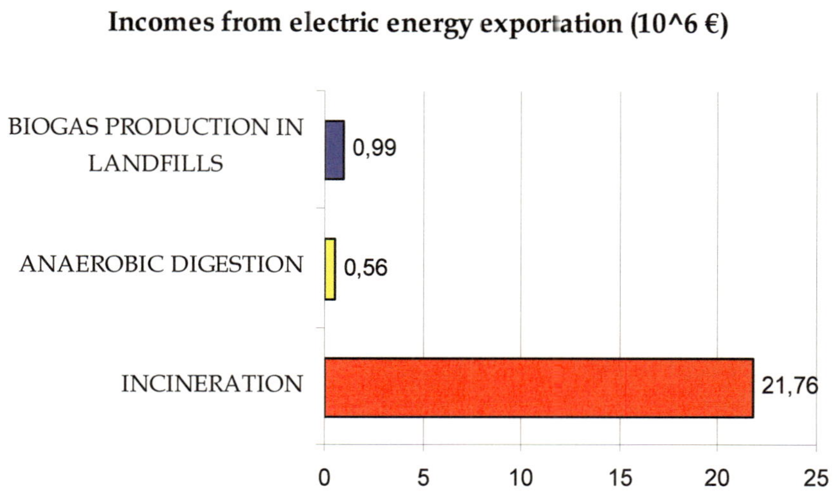

Fig. 12. Revenues proceeding from electric energy exportation.

- The controlled settling in landfills is, definitely, the last step of any integrated solid waste management system and, it's only be applied for wastes that cannot be treated by any other solution;
- In the selection of one or more solid waste treatment techniques and, looking for factors like, economic reality of the country, quantity of waste production and collected, advantages and disadvantages of each kind of treatment method, etc., the technique most recommendable for the treatment of dangerous solid wastes in Portugal, and, generally, in the 27's EU is the co-incineration in cement furnaces, because this option implies less investment costs when compared with the pyrolysis by plasma with vitrification (PPV). Although, it's sufficient adapt them to the gaseous emissions treatment;
- In the co-incineration on cement furnaces, the burn process is particularly controlled, since emits less pollutants atmosphere emissions than the dedicated incineration;
- The co-incineration brings also economic advantages through fuel savings, regarding that 60% of cement cost production are applied to the energy consume;
- However, there is a bigger restriction in the waste admittance on the cement furnaces than in the dedicated incinerators, because, in that ones, occurs, simultaneously, the clínquer fabrication process, which could be severe affected through the furnace inlet of undesirable materials;
- One big advantage of any solid waste management integrated system consists in the valorisation of co-products, such as, the ashes and slag's from the incineration process;
- It's in the incineration that the electricity production is substantially bigger, followed by the biogas energetic valorisation in landfills and, finally, followed by the anaerobic digestion. This fact is extremely important in order to implementate an energetic valorisation in a dedicated incinerator;
- The system PPV is a treatment method that can be applied to any kind of solid waste, including the dangerous ones that could not be incinerated;
- None technology is 100% clean, but the PPV system is, far away, the treatment method which less negative environmental impacts creates, which is a very important advantage under a environmental and social perspectives.

8. Acknowledgments

The authors are grateful for all contributions that the following entities are given to the formulation of this chapter: Valorsul's Integrated Management System of urban solid waste (www.valorsul.pt), Lipor's Integrated Management System (www.lipor.pt) and Amarsul's Integrated Management System.

9. References

Brunner, C. (1994). *Hazardous Waste Incineration*, 2nd edition. McGraw Hill, NY, USA

Camacho, S. (1995). *Mixed waste disposed and energy recycling by plasma pyrolysis/vitrification.* Seoul National University Editions, Seoul, South Korea

Cocarta, D.; Ragazzi, M.; Badea, A. & Apostol, T. (2007). Critical analysis of the EU regulation on waste incineration plants related to heavy metals emissions,

Proceedings of Sardinia 2007, Eleventh International Waste Management and Landfill Symposium, October 2007, IWWG, Cagliari, Italy

Damgaard, A.; Riber, C.; Hulgaard, T. & Christensen, T. H. (2007). Life-cycle assessment of waste incinerators - the significance of increasing air pollution control on the environmental, *Proceedings of Sardinia 2007, Eleventh International Waste Management and Landfill Symposium*, October 2007, IWWG, Cagliari, Italy

Formosinho, S.; Pio, C.; Barros, H. & Cavalheiro, J. (2000). *Dangerous industrial wastes treatment report*. Independent Scientific Council of Co-incineration environmental inspection and control, Aveiro, Portugal

Freeman, H. (1988). *Standard Handbook of Hazardous Waste Treatment and Disposal*, McGraw-Hill Book Company, NY, USA

Higgins, T. (1989). *Hazardous Waste Minimization Handbook*, Lewis Publishers, NY, USA

Lapa, N. & Oliveira, J. (2002). An ecotoxic risk assessment of residue materials produced by the plasma pyrolysis/vitrification (PP/V) process. *Waste Management*, Vol. 22, No. 3, (June 2002) (335-342), ISSN 0956-053X

Levy, J. & Cabeças, A. (2006). *Resíduos Sólidos Urbanos, Princípios e Processos*, AEPSA, ISBN 989-95059-0-0, Lisboa, Portugal

Odum, E. (2001). *Ecology Fundamentals*, 6th edition, Fundação Calouste Gulbenkian, ISBN 972-31-0158-X, Lisboa, Portugal

Oliveira, J. (2000). *The paradigm/paradox of solid waste valorisation: from matter to energy, from useless to essential.* FCT/UNL Editions, Almada, Portugal

Oliveira, J. (2005). *Environmental Management*, Lidel, ISBN 972-757-328-2, Lisboa, Portugal

Partidário, M. & Jesus, J. (1994). *Environmental Impact Evaluation*, CEPGA, ISBN 972-96010-0-0, Lisboa, Portugal

Peavy, H. S.; Rowne, D. R. & Tchobanoglous, G. (1985). *Environmental Engineering*, McGraw-Hill Inc., ISBN 0-07-049134-8, USA

Pio, C.; Barros, H.; Cavalheiro, J.; Dias, R. & Formosinho, S. (2003). *Co-incineration, a war to eight o'clock news*, Campo das Letras, ISBN 972-610-654-0, Porto, Portugal

Puna, J. (2002). Which solutions for the treatment of Industrial Solid Wastes? *Proceedings of Scientific and Technological Conference in Engineering*, pp. 6-19, Lisboa, May 2002, ISEL, Lisboa, Portugal

Puna, J. (2005). Municipal Solid Waste Integrated Management. *Proceedings of Engineering and Municipal Management National Meeting*, Porto, June 2005, CENERTEC, Porto, Portugal

Puna, J. & Baptista, B. (2008). The urban solid waste integrated management - environmental and economic-energetic perspectives, Vol. 31, No. 3, (April 2008) (645-654), ISSN 1678-7064

Russo, M. (2005). *Solid waste treatment*. University of Coimbra Editions, Coimbra, Portugal

Scoreco (1997). *The co-incineration of industrial wastes in cement furnaces – technical document*. Scoreco, Lisboa, Portugal

Tchobanoglous, G.; Theisen, H. & Vigil, S. A. (1993). *Integrated Solid Waste Management*, McGraw-Hill International Editions, ISBN 0-07-063237-5, Singapore

Wathern, P. (2000). *Environmental Impact Assessment, Theory and Practice*, Routledge, NY, USA

9

Managing Wastes in Asia: Looking at the Perspectives of China, Mongolia and the Philippines

Kevin Roy Serrona, Jeong-soo Yu and Jia Che
Department of Interregional Environmental System,
Graduate School of International Cultural Studies, Tohoku University
Sendai City,
Japan

1. Introduction

The state of solid wastes in Asia reflects a situation where poverty and inefficient resource management are intertwined. Urban centers continue to grow and so with population. Consumption of resources necessarily goes up and generation of wastes is increasing at an alarming rate. Consequently, methane and carbon dioxide emission are rising. In the global landscape, greenhouse gas emission is being felt strongly with the melting of ice in the North Pole, changing seasonal patterns, and the imminent threat to the submersion of small islands. The effects of climate change as a result of man-made activities threaten everyone across social classes and geographical location. But developing countries are more exposed to vulnerability in view of poor resources and technology to cope with it. Poor planning, limited financial capacity, lack of technical know-how and toothless laws are some of the barriers that do not permit them to implement environmentally-sound, economically-viable and socially acceptable waste management programs. Among the serious problems needing serious attention is what to do with end-of-life (ELV) or used vehicles and the accumulation of non-biodegradable wastes like plastic that are left on the streets, drainages and water bodies.

A number of countries in Asia have jumpstarted the campaign to reverse the problem of ELV accumulation. The European Union (EU) pioneered an ELV law in September 2000. Japan and Korea followed suit with the former passing an Automobile Recycling Law in January 2005. Korea, on the other hand, passed the Resources Recycling Law in January 2008. These countries recognized that a distinct ELV law is necessary within the framework of the extended producer responsibility (EPR) system. An international cooperation is being pursued by the Japanese government in partnership with Tohoku University and car manufacturers like Hyundai, Kia Motors, Shanghai GM and Volkswagen through the Asian Environment-friendly Automobile Forum to promote knowledge and awareness on ELV recycling in Asia. Experts from the academe, government and the private sector converge annually to exchange ideas and technical know-how on how to best address accumulated used vehicles and assist in crafting policies that will mandate recycling. To realize this, the first forum was held in China in 2007 and in Korea in 2008. This year, it will be held at Tohoku University in Sendai, Japan.

In another front, unregulated municipal wastes have led to tremendous problems like air, water and land pollution. Recycling in developing countries is poorly implemented. In fact, the absence of a mainstream recycling system has led to the emergence of an informal sector consisting of wastepickers and small junkshops who recover wastes at disposal sites. Lack of jobs and poverty are the main reasons for this scenario. However, small communities have been creative and proactive to address the problem. In Manila, Philippines, a women-led and community-based recycling project is being implemented where used plastics are converted into handicrafts and are exported and sold locally. Most important aspect of this undertaking is that women gained jobs and are able to contribute to the family income. The project has empowered them through economic and social means and the community benefitted from this environmental initiative.

This paper discusses the state of ELVs in China and Mongolia with reference to the experience of Japan and Korea as far as car recycling is concerned. In addition, a community-based plastic recycling initiative in Metro Manila, Philippines is discussed and analyzed. The marriage of the above in this paper points to the importance of highlighting recovery of usable ELV parts and reducing plastic wastes in the municipal solid waste stream in the case of developing countries.

2. End-of-life vehicle recycling: the forerunner

The EU first passed a law on ELV recycling in 2000 which became the precursor for other countries like Japan and Korea to implement similar legislation. Europe has witnessed an exponential increase in the number of vehicles produced starting from the 1990s. About 14.5 million cars per year have been manufactured since 1998 with about 17 million in 2002. The impact that the industry created to the environment is huge in terms of energy and resource consumption, hazardous emissions, waste generation of toxic substances and disposal. It is estimated that about 75% of ELV in EU are recyclables while the remaining 25% are disposed of in landfills (Kanari et al., 2003). This prompted the passage of a recycling law that caters to ELVs in Europe.

In Japan, the "Law for the Recycling of End-of-life Vehicles" was implemented in 2005. The main feature of the law is that automobile manufacturers and importers have the responsibility to collect and recycle air bags and shredder residues generated during the treatment process of ELVs. End users, on the other hand, pay the appropriate recycling fee for car owners during the first car inspection.

Korea passed into law the "Act for Resource Recycling of Electrical and Electronic Equipment and Vehicles" in April 2007. The main purpose is restrictions on the use of hazardous substances and manufacture of products that facilitates recycling. A manufacturer or importer is required to develop recycling technology and provide technical support to vehicle scrapping business and dismantled recycling business. The target recycling rate is 85% by 2014 and 95% by 2015 with an energy recovery of not more than 5% by the former and no more than 10% by the latter.

Overall, the ELV recycling laws mentioned above can be summarized as follows:

	European Union	Japan	Korea
Timeframe	September 2000	January 2005	April 2007
Vehicle covered	Passenger cars with seating capacity of nine or less and commercial vehicles with gross value weight of 3.5 tons or less	Four-wheeled passenger cars and commercial vehicles	Passenger cars with seating capacity of 9 or less and trucks with maximum weight of 3.5 tonnes
Center of responsibility	Manufacturer e.g. establishment of ELV collection and recycling network	End user e.g. surrender of ELV and payment of necessary fees	Manufacturer e.g. development of recycling technology and technical support to vehicle scrapping business
Costs	Borne by the manufacturer	Borne by end users through a fund management corporation	Borne by the manufacturer (zero cost to the end user)
Information system	Monitoring and inventory of ELV samples are done. Same with Korea.	Monitoring focuses on airbag, freon gas and automobile shredder residue (ASR) only	Every ELV is checked including weight and type, etc.

The ELV laws passed by Japan and Korea have significant impacts on Asian countries. Skyrocketing prices of scrap irons, global warming and cross-border shipment of waste are some of the factors that are shaping up the automobile recycling industry. Vehicle recycling has revolutionized recycling technologies and fueled economic gains. At the same time, it has also uncovered the social aspect of automobile recycling. In developing countries, poor people are engaged in the recovery of metals and used automobile parts and a source of profitable income for small-scale used car dealers.

3. Current situation of ELVs in China and Mongolia

3.1 China today

China is, undoubtedly, a fast growing economy in the world. At the same time, it has also overtaken the United States as the largest emitter of greenhouse gas which is casting a serious shadow under the ray of global warming. The figures are staggering as far as the volume of vehicles and ELVs are concerned in China today. It was projected that by the end of 2006, the volume of vehicles running on the road had reached 32 million while that of ELVs were more than 1.5 million by the end of 2005. In 2003, the volume was 23.82 million while ELVs were 3-5% of the total. China passed a law regulating the disposal and recycling of ELVs in 2001, a year after the EU made its own ELV law. However, progress has been slow as far as the rate of dismantling is concerned - only 10% at the onset of 2004 (Chen, 2005).

The main feature of the law dubbed as "Statute 307" is the declaration of a vehicle as ELV based on some technical specifications like mileage accumulation and service rendered in years. For example, a passenger vehicle with a mileage of up to 500,000 kilometers classifies it as an ELV. In China, commercial vehicles comprise the most number of ELVs while personal cars are only a small portion. Obsolete imported cars are also increasing which came from the US, Germany, Japan, Korea and the rest from other countries. These used vehicles are usually dismantled for their valuable metal parts. In addition, the law stipulates that vehicle owners sell their ELVs to a vehicle recycling enterprise. Sale to unregistered or unqualified individuals and even donations are not allowed.

The law also requires the establishment of an information system to monitor, manage and administer the entire vehicle life cycle from design to dismantling and recycling. In this regard, local administrative districts are the focal point in partnership with auto manufacturers. To further strengthen the law, a supplementary regulation was issued in 2006 called the "Motor Vehicle Product Recovery Technology Policy" in which one of the salient features is that manufacturers together with material and equipment manufacturers take the responsibility of sustainable recycling. It requires vehicle producers to work in tandem with operators doing component production, dismantling, remanufacture and recovery of ELVs and material recycling. The target is, by 2010, vehicle producers and agents of imported vehicles will be responsible for the recovery and treatment of their vehicles with the necessary fees involved. And they shall establish tie-up with enterprises involved in the dismantling and shredding of ELVs by providing technical information e.g. vehicle dismantling manual, etc.. In other words, the whole gamut of a vehicle will be considered to facilitate dismantling and recycling.

Imported vehicles account for a large volume of cars in China. A breakdown of countries importing vehicles to China is shown below:

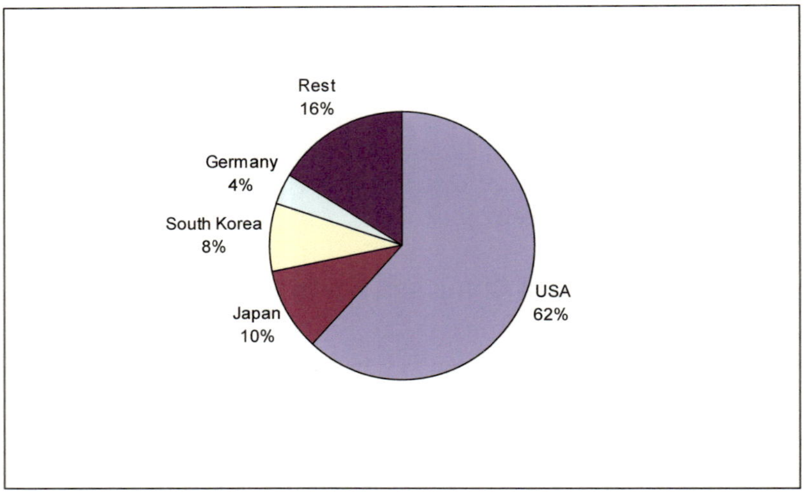

Fig. 1. Imported vehicles in China

The United States imports the most number of vehicles followed by Japan, South Korea and Germany. The rest represents various origins. Large importation of cars occurred in the 1990s with passenger cars accounting for the biggest proportion. In 2000, for example, it constituted 51%. Metal parts of imported ELVs are the ones recycled. On the other hand, table 1 shows sales, car possession shows sales, car possession and used car generation for the period 2005-2007:

Year	Sales	Car possession	Used car	Rate of used car generation (%)
2007	8.88 million	43.58 million	1.97 million	4.5%
2006	7.16 million	36.97 million	1.79 million	4.8%
2005	5.7 million	31.6 million	1.05 million	3.3%

Table 1. Vehicle sales and used car generation in China

Looking at car production, there has been a steady increase in the volume of cars manufactured in China representing different automakers as shown below:

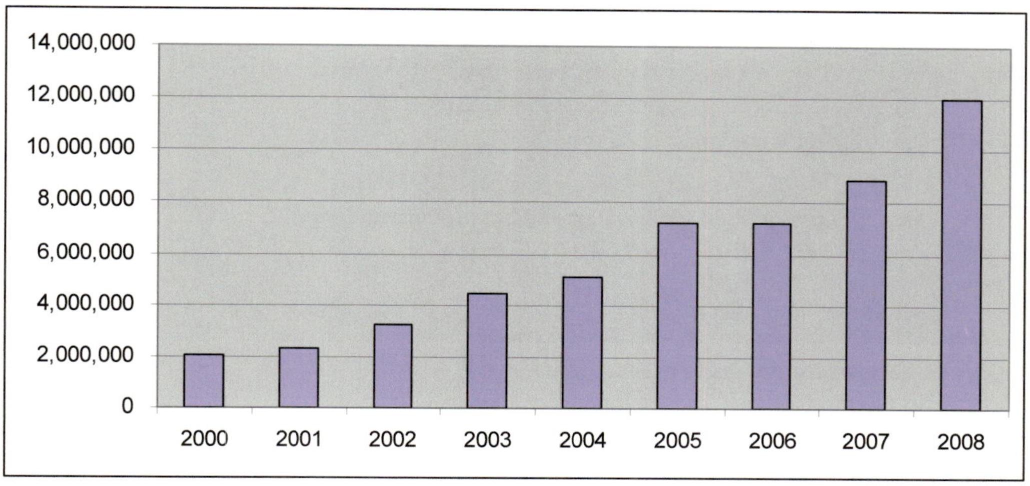

Fig. 2. Volume of car production in China (2000-2008)

ELV recycling in China faces many challenges. Lack of professional equipment, low dismantling efficiency, low recycling rate and environmental pollution are some of the issues confronting the industry. The dismantling aspect is described as "manually-based" due to the low cost of labor. In addition, they are outdated and environmental measures are poor. A situation too far when compared with the dismantlement technology in Japan. The roads are not paved and in the factory sites, wasted oil and fluids are left dripping into the ground. Fluorocarbons that pollute the air are neither collected. Iron and metallic resources are dismantled manually. It is said that the amount of automobile shredder residue (ASR) generation is near zero in China due to the manual dismantling scheme. An example of a dismantled car in Shanghai City is shown in figure 3.

Shanghai City was made as a pilot industrial demonstration of ELV dismantling and disposal in 2005 by virtue of Stature 307, the law which regulates the disposal and recycling of ELVs. The objective is to disassemble used commercial vehicles to be used as spare parts and recycle rubber, plastic and metal materials. The overall goal is to "establish an ELV recycling engineering system and remold the ELV recycling industry from an extensive to intensive and environmentally benign industry." Based on the initiative, significant achievements were made in terms of metal retrieval from ELVs as shown in table 2.

Fig. 3. Dismantled used vehicle in Shanghai City

Year	In-Use Vehicles	ELVs	Rate of ELVs In Use	Volume of Metal Recycling (tones)	Ferrous Metal (tones)	Nonferrous Metal (tones)
1995	308,258	9,171	2.98%	23,315.12	22,949.06	366.06
1996	343,815	8,630	2.51%	17,805.24	17,393.14	412.10
1997	387,538	11,194	2.89%	23,593.35	23,112.54	480.81
1998	404,491	13,783	3.41%	28,958.14	28,440.20	517.94
1999	451,419	11,774	2.61%	24,706.55	24,233.99	472.56
2000	492,025	11,119	2.26%	20,517.14	20,086.86	430.28
2001	518,693	13,773	2.66%	21,185.07	20,836.97	348.10

Table 2. Volume of in-use and end-of-life vehicles in Shanghai Administrative District and metals reclaimed from ELVs

3.2 Mongolia

Mongolia is a country in progress. It has vast natural resources but population is relatively small. As of 2009, its population is about 2.6 million (World Bank, 2009). About 61% or 1.58 million are living in urban areas. Ulaanbaatar, the capital city, accounts for the majority of the urban population estimated at 994,000. As such, motor vehicle possession is also concentrated in urban areas. In the capital city, car ownership rose from 28,119 in 1995 to 104,539 in December 2007. The origins of these vehicles vary as shown in figure 4.

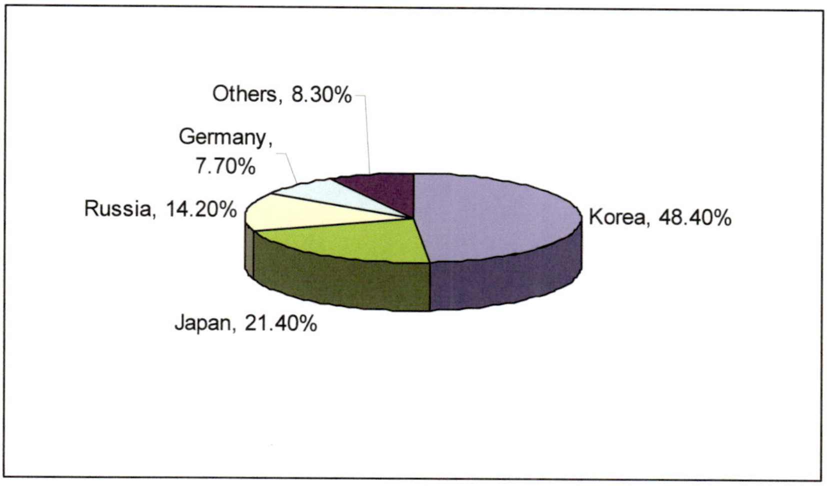

Fig. 4. Breakdown of countries exporting cars to Mongolia

Korea is the top exporter of cars to Mongolia with almost 50% of vehicles followed by Japan then Russia. European cars constitute a small portion of cars in Mongolia. A detailed breakdown of vehicles coming from various countries is shown below:

	1995	**1996**	**1997**	**1998**	**1999**	**2000**	**2001**	**2002**	**2003**	**2004**	**2005**
Russia	3914	1393	290	2288	3025	6190	2370	1944	1086	541	448
China	77	137	114	28	27	23	-	5	8	14	21
Germany	79	215	273	588	291	564	456	333	182	133	142
Korea	472	607	455	1910	737	1752	2512	1064	2230	3080	3227
Japan	110	421	493	765	716	2808	2747	3717	6666	8987	9289
Others	1558	223	112	111	66	172	224	124	150	178	195
Total	**6210**	**2996**	**1737**	**5690**	**4862**	**11509**	**8309**	**7187**	**10322**	**12933**	**13322**

Table 3. Countries importing cars to Mongolia

The above table shows the increase in the volume of vehicles from 1995 up to 2005. From 6,210 vehicles in 1995, the total number of vehicles in 2005 was 13,322. A study made by the Asian Development Bank (ADB) showed that as of December 2007, a total of 196,332 vehicles were registered nationwide (ADB 2008). Out of this, eighty percent (80%) of vehicles inspected did not pass national or international emission standards. The study further revealed that more than 50% are over 11 years old and 30% are 7-10 years old. This situation has worsened the ambient air in Mongolia with pollutants being emitted by old vehicles.

ELVs in Mongolia, therefore, abound with used cars still on the road. The absence of technology to recycle is one factor for the large volume of ELVs. Figure 5 reflects the age of vehicles in Ulaanbaatar where the largest concentration of vehicles are located.

Mongolia does not have a legislation on ELV recycling and as a result, used vehicles accumulate. There is manual recovery and sale of used parts but there is no recycling in the absence of recycling technologies.

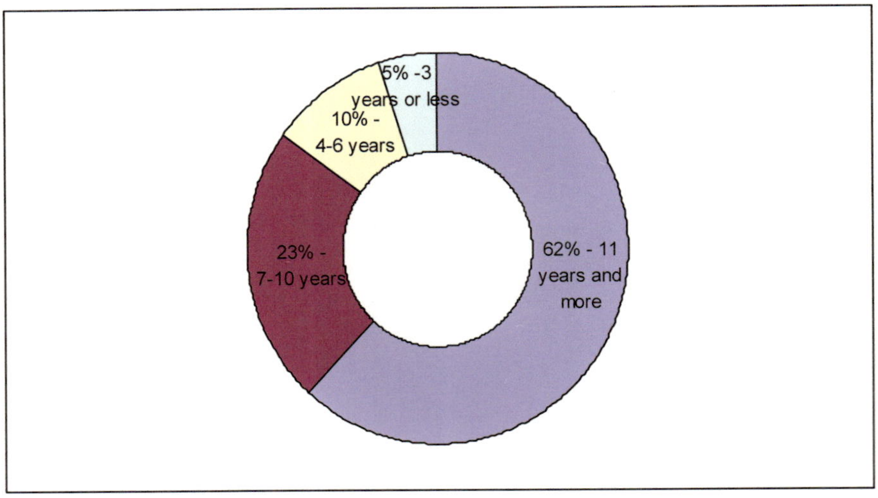

Fig. 5. Vehicle usage by years in Ulaanbaatar, Mongolia

Fig. 6. ELV in Mongolia

4. ELV dismantling: Efficiency and costs

The drive towards ELV recycling resulted in two (2) methods, namely: manual and machine-based dismantling. In China, where manual dismantling is usually involved, a comparison was made between the two. Figure 7 shows that machine dismantling results in more weight than manual dismantling.

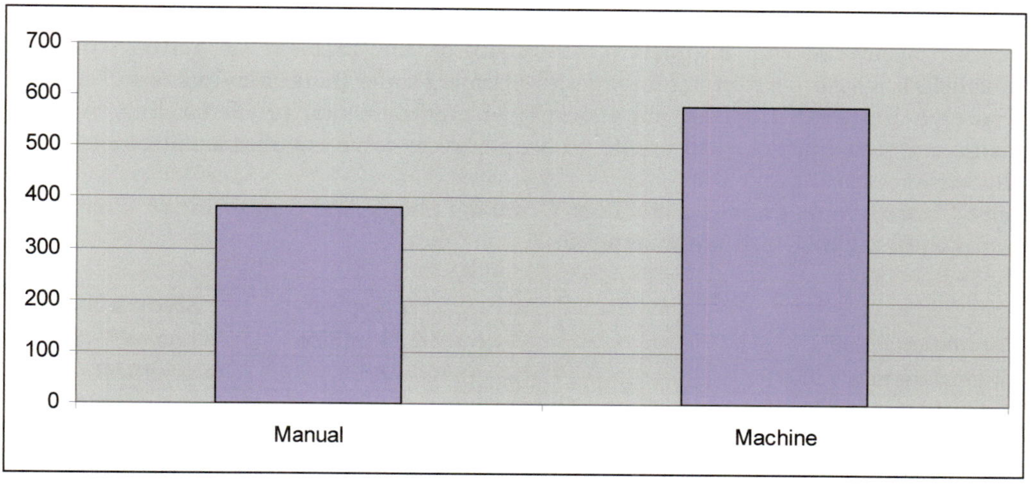

Fig. 7. Weight of pressed block from a dismantled ELV (kg)

Manual dismantling takes longer time than machine dismantling. However, more valuable parts are recovered in the former which translates into more parts to be sold or recycled. In the latter, there is more waste since the machine destroys some useful parts. On the other hand, the figure below shows a comparison in terms of the value of recovered parts using both manual and machine-based methods:

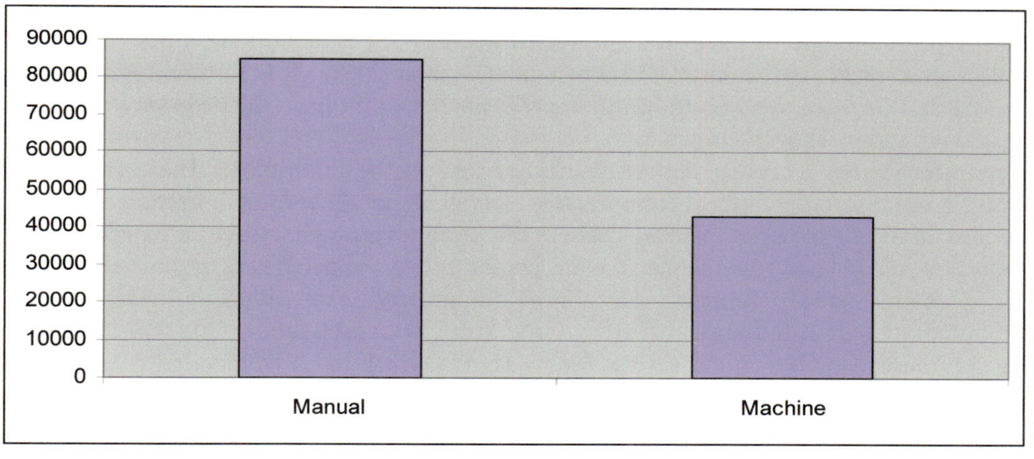

Fig. 8. Value in terms of ELV recovered parts in Japanese Yen (million)

As explained earlier, manual dismantling recovers more useful parts than machine mechanical dismantling. This translates into more monetary value for recovered parts.

5. Community-based recycling in the Philippines: From small to big steps

The state of solid waste management in the Philippines, particularly, in urban areas like Metro Manila is a microcosm of the prevailing situation in developing countries. Problems abound such as unregulated disposal, poorly maintained disposal sites, prevalence of informal waste recovery and low recycling rate. These are complemented with lack of data on volume of waste, generation rate and composition of waste. This situation prevents a scientific approach and a practical resolution of the problem of garbage disposal. Nonetheless, a basic tenet in community development states that communities will continue to survive and evolve amidst the enormity of environmental problems. Initiatives will emerge and best practices will unfold as people try to solve environmental problems like solid waste.

Metro Manila is a metropolis with about 12 million people and comprising of 17 cities and municipalities. Quick facts are shown below:

	Philippines	**Metro Manila**
Population (2009)	92.23 million	11,553,427 (13%)
Unemployment (2009)	7.5%	13.5%
Annual per capita	US$330	US$6,827
Average family income (2006)	US$3,798	US$6,827
Literacy rate (2003)	84%	94.6%
Poverty incidence	32.9%	10.4%

Table 4. Quick facts about the Philippines

Unemployment rate in Metro Manila is high at 13.5%. Poverty incidence is likewise soaring with 33% nationwide and 10.4% in the Metropolis. It can be said that poverty is related to environmental problems like solid waste. People who do not have or cannot find a job resort to wastepicking to survive. Wastepickers are present in almost all the major disposal sites in Metro Manila. In the absence of a formal recycling system e.g. EPR, informal recycling thrives. But the plight of those in the informal waste sector is deplorable considering that a typical wastepicker earns only US$2-3 per day (Serrona, 2009). It is further aggravated by the fact that they are exposed to health hazards as they don't have the necessary protection e.g. gloves and mask while at work.

Community-based recycling initiatives are not new in the Philippines. There are various people's organizations in local communities as well as non-government organizations that are into advocacy of certain issues. There is the local government structure which plays an important role in local governance. In some communities, the people are organized by sector e.g. women, youth, farmers, etc.. The community exemplifies a heterogeneous conglomeration of individuals and groups interacting with each other. One typical example is a community in Ugong, Pasig City, Metro Manila which has a women-led community - based recycling project. The name of the organization is KILUS or "Kababaihang Iisa ang Layunin para Umunland ang Sambayanan" (Women Who Are United for the Progress of Society). Formed in August 1997, the group was then called "Samahan ng mga Kababaihan ng Ugong" or Women's Group of Ugong. Its vision was for the cleanliness of the community. It became the partner of the local government in solid waste management. It was a partnership which earned the title "Cleanest Barangay along Pasig, Marikina and San Juan River. Then, it expanded and was formally registered in 1999 as KILUS. Presently, it is composed of 500 women members.

The path that KILUS took was unique. It focused on livelihood opportunities from garbage. As it evolved, the organization underwent skills training on handicraft making. They discovered a plastic material called "doy pack" which is sourced from a discarded juice container popular in schools and social gatherings. Anticipating a good business out of this material, the group pursued to find creativity in the material. Handicrafts such as bags, home furnishings, footwear, and fashion accessories came into existence as a result of product development. With funding support (loan) from the government, it purchased sewing machines and other equipment for the members to use. The doy packs are sourced from within and outside the community by designated members and are bought at US$0.10 apiece. Some are also collected from manufacturers of doy packs which are usually "rejects." With expanding members and services, KILUS has metamorphosed into a multi-purpose environmental cooperative producing handicraft products and promoting cooperativism as a social vehicle to empower women. The bulk of its products are sold abroad like Japan, USA, Canada, Germany, Los Angeles, and London. The organization has an interesting program for its members like conflict-resolution meetings and the absence of vertical bureaucracy. Any member can directly talk to the manager and vice-versa. This makes KILUS an informal group but with an effective communication strategy among members. The result is smooth leadership and easy resolution of conflicts.

6. From doy packs into handicrafts: Sustaining lives

The process of converting used doy pack into handicrafts involves low-technology and is labor-intensive. KILUS is not really intent on modernizing its process as it exists on the creativity and dedication of its workers. Replacing people with equipment would just translate into loss of jobs and security of its members. It has four (4) product categories:

Products	Items	Price
1. Bags	Shoulder bag	
2. Home furnishings	Book shelf, fruit tray, placemat, lunch box	
3. Fashion accessories	Belt, beltbag, jewelry box, beads	Price ranges from US$5-19
4. Footwear	Sandals	

Table 5. KILUS products by category

The above reflects the experiments that the organization did to suit the needs of its customers. Based on the data, sixty percent (60%) of its shipment abroad consists of bags. On the other hand, the process of creating handicrafts requires major steps.

The first step involves the collection of doy packs from schools, factories, funeral parlors and other establishments. Each used doy pack is brought at US$0.002. Those coming from factories which are considered "rejects" are bought by KILUS at US$0.17 per kilo. Factory rejects require less time and effort to clean compared with those coming from schools which have to be washed up. Production is done at KILUS office and at the home of its workers. This arrangement is to give them time to take care of household chores and at the same time work on their assignment. Tasks done at home are those associated with washing, strip-making and beads-making. Once finished, they go to KILUS to submit their outputs. These are then recorded for payment purposes.

Quality control comes next where necessary corrections are made. Interestingly, quality control is done using only a lamp and a scissor to remove imperfections. After quality control, an inventory of finished products is made and preparation for packaging and subsequent shipment is made. The organization makes sure that they meet the timetable

Fig. 9. A bag made out of doy pack (inset photo)

requirements of their customers. Shipment delays are tantamount to losing loyal buyers. Air and sea freight are used as mode of shipment with the former taking one (1) week to deliver while the latter takes a month. International buyers have to pay a down payment fee of 50% upon ordering. There is no established peak and off-peak seasons as demand depends on the buyers.

Orders from abroad serve as the lifeblood of KILUS operations. In this regard, it ensures that quality products are produced. For the workers, creating a single handicraft is all about pouring their hearts into it. The production stage might be repetitive but they treat every single product as unique that will be cherished by customers.

Markets and sales

KILUS started selling products in 2002. It earned its customers through networking and showcasing its products in product exhibitions. Over the years, countries which have patronized its products are shown in the following figure:

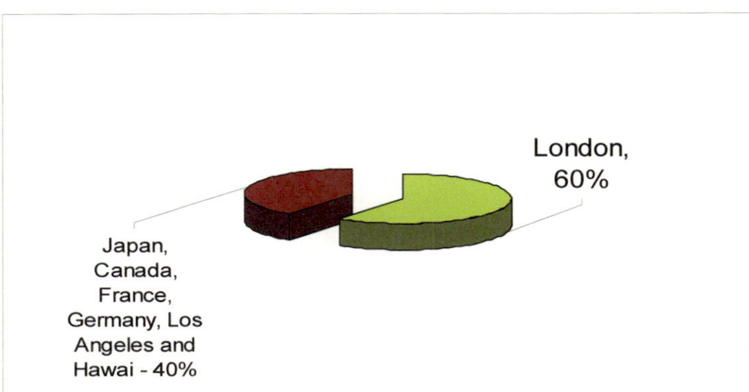

Fig. 10. KILUS International Markets

It will be noted that London is the main market of KILUS with 60% of its market. This can be attributed to the increasing consciousness among Europeans to recycle waste and be an advocate of an environmentally-sound lifestyle. Japan, Canada, France, Germany, Los Angeles and Hawaii comprise 40% of the market. Locally, it has one (1) shop in a shopping mall. To reach out to more local customers, it participates in bazaars. Still, international buyers dominate their sales. The sale of KILUS from their international buyers is shown below:

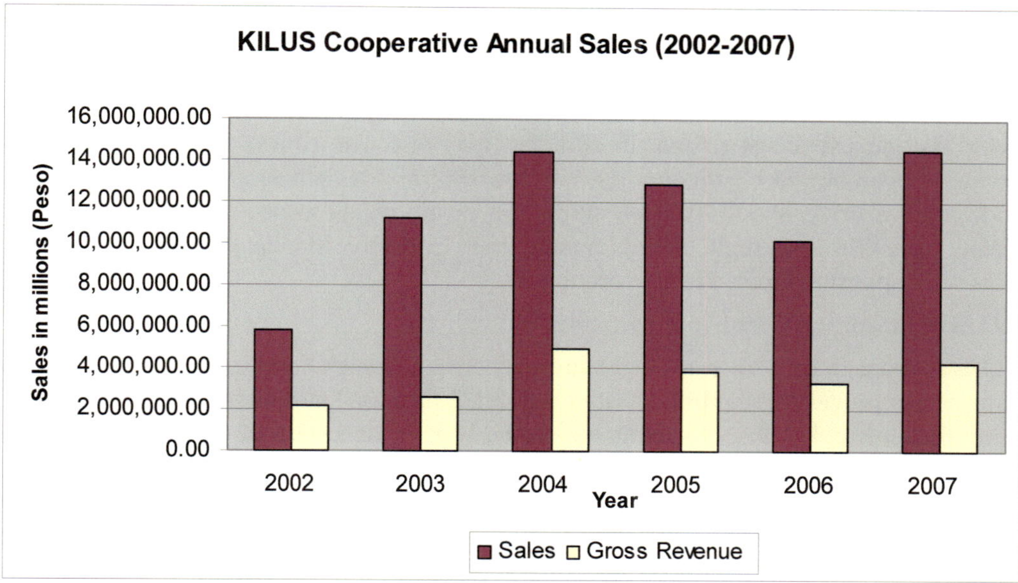

Fig. 11. KILUS annual sales

The above figure is indicative of the profitability of KILUS products. In 2002, sale reached approximately US$126,000. Subsequent sales went up in 2003 and 2004 and a slight decline in 2005 and 2006. However, sales went up again in 2007. The gross revenue reflects the net amount as a result of payments made to overhead costs and salaries for KILUS workers.

Salaries and benefits

Women workers are employed by the Cooperative. Payment is made based on the type of work that one does. As stated earlier, there are office-based and there are home-based workers. Home-based workers are paid apiece and are required to remit their outputs to the KILUS office on assigned dates. Submission of outputs is strict as international buyers set deadlines for shipment. What sets apart KILUS from other enterprises is there is no age limit if one desires to work. In an interview with Ms. Carmelita Elec, Business Manager of KILUS, as to the reason why women workers are preferred than men, she said that women are easy to talk with. "We can talk with women without a bottle of beer unlike men," according to Ms. Elec. Firm decision-makers are what characterize women and that is why the Cooperative hires them.

In terms of salaries, the following table shows the compensation of workers based on tasks:

Job category	Income
Sewers	US$278/month
Designers	US$204/month
Production manager	US$204/month
Raw materials	US$139/month
Receiving	US$93/month
Quality control	US$5/day
Cutter and sorter	US$3/2,500 pieces
Washer	US$3.2,500 pieces

Table 6. Jobs and salaries at KILUS

KILUS is not an ordinary enterprise. On top of their regular salaries as stated above, members also get benefits like rice, health insurance, groceries, free medical check-up, dividends and patronage refund. In addition, they also can avail of marriage counseling services. As an outreach program, the Cooperative provides scholarship to the children of workers numbering about 20. Each student is given an allowance of US$25 for school-related use. With the high unemployment rate in Metro Manila, KILUS provides the necessary support to women in the community.

KILUS replicability: Cost-benefit analysis

KILUS appears to be a boon to the community from the social and economic standpoints. However, to prove this requires a cost-benefit analysis (CBA) as a tool for decision-making. To do this, a simple CBA was made where the benefits and costs of operating KILUS are made from the perspectives of Metro Manila and the host community. There are two scenarios: with KILUS and without KILUS project. In the analysis, the net present value (NPV) which pertains to the benefits less costs was determined. Based on the study made by the Asian Development Bank (ADB), Pasig City where the Cooperative operates, spends about US3,611 on municipal waste management expenses which is nine percent (9%) of the total local government expenditures. This is lower compared to the average expenditures of a local government unit (LGU) in Metro Manila which is US$3,558,345 or 13% of the total LGU expenses.

The table above reveals that KILUS is beneficial for replication. The first two columns shows that benefits exceed costs. The net social benefits are higher with the project and significantly lower without the same project. On the fourth column (Pasig City perspective), the net social benefit is negative. It is worth to note that a project is worth recommending if the NPV is positive. As such, a negative NPV like the one in the fourth column (US$-511) is not worth recommending. Based on the above, KILUS is beneficial for the community from the perspective of providing jobs, reducing SWM costs and in cleaning up the environment.

7. Local government savings from recycling

The operation of KILUS directly benefits the LGU in terms of reduced SWM costs. Less doy packs on the streets and in street canals means less expenses for the LGU in terms of collection, transportation and disposal. To quantify the savings, a CBA was made to determine the savings of Pasig City in terms of tipping fees collected from the residents and subsidies from the government with doy pack recycling.

Impacts[a]	With KILUS		Without KILUS	
Project benefits:	Metro Manila perspective	Pasig City perspective	Metro Manila perspective	Pasig City perspective
Revenue from sales	270,216[b]	270,216	0	0
Reduced SWM cost[c]	75,739[d]	377.52	76,116.77	3,432
Reduced transportation cost to landfills	3,720	85.8	3,806	171.6
Livelihood for the poor/unemployed	116	116	0	0
Less doy pack on streets and elsewhere	76,117	343	76,117	343
TOTAL BENEFITS	425,908	271,138	156,039.77	3,947
Project Costs:				
Construction costs[e]	42,782	42,782	0	0
Operational costs (e.g.salaries, utilities, maintenance, etc.)	75,022	75,022	0	0
Per capita SWM cost	8	6	8	8
Waste disposal costs	74,836.77	$1,280	76,116.77	3,450
TOTAL COSTS	192,648.77	1,286	76,124.77	4,458
Net social benefits = NPV	**233,259**	**269,852**	**79,915**	**-511**

[a] refer to inputs (required resources and outputs)
[b] average KILUS sales
[c] obtained by multiplying the annual SWM cost (US$3,432) of Pasig City by 11% which is the percentage of used doy packs
[d] obtained by subtracting the total SWM costs of LGUs in Metro Manila from US$377.52 which is the amount reduced in SWM cost by Pasig City with daypack recycling
[e] assumes that space is provided by the local government but construction costs are shouldered by the project

Table 7. Cost-benefit analysis for KILUS project (in US$)

	2002	2003	2004	2005	2006	2007
Operating costs*	**75,021.73**	75,021.73	75,021.73	**64,564**	**75,940**	**84,561**
Sales**	**270,322.74**	270,322.74	270,322.74	**277,677**	**219,960**	**312,690**
Average sale price***	**5**	5	7	**7**	**11**	**11**
Gross revenue from sales	**81,965**	81,965	81,965	**82,366**	**70,754**	**92,725**
Per capita SWM (US$)****	8	8	8	8	8	8
Benefits	**10,246**	10,246	10,246	**10,295**	**8,844**	**11,590**

* 2002, 2003 and 2004 were assumed average costs based on 2005, 2006 and 2007 available figures
** 2002, 2003 and 2004 were assumed average sales based on 2005, 2006 and 2007 available figures
*** price of KILUS products range from US$5 to US$18 based on 2007 data
**** Average per capita SWM cost in Metro Manila (ADB)

Table 8. Cost-benefit analysis of KILUS operations (in US$)

The above table shows the benefits in US dollars from KILUS operations which were obtained by dividing the gross revenue from sales over the tipping fee which is US$8/ton. The tipping fee refers to the disposal fee where Pasig City dumps its waste. With doy pack recycling, the average yearly savings of the City amounts to US$10,244 as against the US$3,450 yearly expenses based on 2001 figures (Westfall & Allen, 2004). This is a significant savings considering that the per capita contribution to SWM expenses is only US$0.50 which is merely six percent (6%) of the total SWM expenses of the City. The recycling project allows the City government to save money and channel the savings to social services needed by the community. The amount of plastic going to landfills is also reduced. Further, the following table shows a comparison of the cost-benefit involved under two scenarios: waste disposal or business-as-usual and doy pack recycling:

	Waste disposal (business-as-usual)	With doy pack recycling
KILUS Income	US$0	US$10,244
LGU expenses (annual)	US$3,450	US$1,280

Table 9. KILUS contribution to reducing LGU SWM expenses

Savings in LGU expenses amounting to US$1,280 was computed by dividing the income of KILUS over US$8 which represents the tipping fee. Baseline figures show that without KILUS, the LGU will be spending US$3,450 annually for SWM related expenses. Moreover, plastic composition is 21% of the total MSW generation in Pasig City based on the ADB study. By weight, daily waste generation by the City is around 273 tons. Plastic constitutes 57 tons out of 273 tons. Assuming that doy pack is 50% of the plastics generated in the City, and KILUS recycles 100% of the doy packs, the reduction in volume is 29 tons. Thus, only 244 toms are left for disposal or further recycling.

8. Success factors

KILUS enterprise shows the feasibility of having community-based recycling at local communities in Metro Manila and elsewhere. The replicability of KILUS depends on a lot of factors such as social and economic situation, political dynamics and the presence of community groups. The role of the latter cannot be disputed since local organizations serve as catalyst for community development. They provide services beyond what the local government could provide such as livelihood and awareness-raising.

In a focus group discussion (FGD) with KILUS workers, five success factors were articulated, namely: good leadership, ability, time, unity and trust (Serrona, 2009). KILUS is championed by vision-oriented leaders and the positive response from the community indicates that when someone sincerely leads, the people will certainly act. The role of the local government is also vital through strong political will and sound legislations.

The FGD also gathered the following needs: product development, more markets and capacity-building for the workers. Product innovation is something that the project needs to embrace because of the evolving preference of its customers. The participants stated that they still lack innovative designs to make products out of doy pack. In this regard, they are coordinating with non-government organizations and other networks for possible trainings on product development. They are also in search for potential partners who can teach them new designs. Market is also a key for their sustainability. On top of their current buyers, they need more to provide more jobs to the needy and more income for the organization.

9. Summary and recommendations

Waste management in Asia poses critical challenges as it grapples with urbanization, increasing population, unregulated greenhouse gas emission and heavy pollution. The threat to the environment is enormous and it goes further to the detriment of people's health and well-being. Actions cannot be deferred as far as reducing GHG is concerned. ELV recycling provides a framework for both developed and developing countries to act together in ending the cycle of pollution that it creates. But there should be equal partnership; one that does not make developing countries the basket for surplus vehicles and parts.

From the economic standpoint, ELV recycling is a sound practice because it meets the demand for scrap iron in the world market. The methodology to dismantle ELV, however, is a tug-of-war between manual and machine. Based on experience in China, more useful parts are recovered from manual dismantling and this translates into money. Cheap labor allows manual dismantling to be sustainable and markets for used parts are always present.

As a recommendation, it is worth to look at ELV and plastic recycling as a boon to the government, local communities and the society in general. The use of local technologies is worth sustaining in light of efficiency and resources recovered as in the case of ELV dismantling and doy pack recycling. Manual recovery helps in providing jobs in developing countries. ELV laws must be put in place in Mongolia and in the Philippines as well to hasten ELV recovery.

In the case of plastic recycling, KILUS utilizes low-technology to make handicrafts. An approach which allows jobs to be generated for housewives and to raise awareness on the need to recover plastic materials. In addition, it allows the promotion of a cooperative where resources and gains are shared by community members. The community spirit of helping together address environmental problems through innovative approach is clearly manifested in the KILUS project. There is valid reason for KILUS to be replicated in other communities citing its social relevance and economic gains.

Future research on ELV may focus on enhancing life-cycle assessment (LCA) of vehicles to ensure that every aspect of it is recovered and there is very little that goes to landfills. Developing countries like Mongolia and the Philippines are rich subjects for research on ELV and plastic recycling. In the case of KILUS, there is a need to do an evaluation impact of the project to assess tangible benefits that the members and the community have received from the initiative. Lessons learned need to be identified as well as best practices in the area of community-based environmental enterprises.

10. References

Akhlaghi, Mehdi et. al (2009). *Little Green Databook*, The World Bank, ISBN 978-0-8213-7849-6. Washington, D.C., USA

Asian Development Bank (October 2008). Mongolia: Urban Development Sector (Rapid Sector Assessment). *ADB Evaluation Report*, Gatti, M. et al, pp. 1-49, ADB Operations Evaluation Department, Manila, Philippines

Che, Jia & Yu, Jeong-soo (2009). Latest Trends on Car Recycling in China and Future Directions. *Proceedings of the Asian Automobile Environmental Forum 2009*, in press, Sendai, Japan. November 13-14, 2009, Sendai

Chen, Ming (2005). End-of-life Vehicle Recycling in China: Now and the Future. *JOM Journal*, Volume 57, No. 10 (October 2005) pp. 20-26

Davaadash, Chagtaltulga; Yu, Jeong-soo & Che, Jia (2006). Environmental Impacts Caused by Vehicle Importation from Japan and Korea to Ulaanbaatar City, Mongolia, *Proceedings of the 18th Annual Conference of the Japan Society of Waste Management Experts*, pp. 14-17, Tsukuba, Japan, November 2006, Japan Society of Waste Management Experts, Tsukuba

Davaadash, Chagtaltulga & Yu, Jeong-soo (2005). Use Car Distribution and Disposal of ELV in Mongolia, *Proceedings of the 16th Annual Conference of the Japan Society of Waste Management Experts*, pp. 228-230, Sendai, Japan, November 2005, Japan Society of Waste Management Experts, Sendai

Kanari, N.; Pineau, L. & Shallari, S. (2003). End-of-Life Vehicle Recycling in the European Union. *JOM Journal*. Volume 55, No. 8 (August 2003) pp. 1-8, http://www.tms.org/pubs/journals/JOM/0308/Kanari-0308.html (accessed October 23, 2009).

Serrona, Kevin Roy (2008). Integrated Urban Waste Management in the Philippines: Interface of Waste-to-Energy and Social Participation. PhD diss., Tohoku University.

Westfall, Matthew S. and Nicholas Allen (2004). *The Garbage Book*. Asian Development Bank: Manila, Philippines

Yu, Jeong-soo (2009). Study on Actual Condition of Extended Producer Responsibility in Japan and Korea: Establishment of a Partnership-based EPR (Japanese), *Proceedings of the Meeting Announcing the Results of Establishing a Recycling-based Society*, pp. 24-27, Fukuoka, Japan, October 27, 2009, Japan Environmental Sanitation Center, Tokyo

Yu, Jeong-soo; Che, Jia & Kosuke, Toshiki (2009). Corrective Strategy of Resource Recovery and Dismantling on the ELV Recycling Process, *Proceedings of the 8th Annual Conference of the Japan Society of Regional Policy*, pp. 61-62, Tokyo, Japan, July 2009, Japan Society of Regional Policy, Takasaki

10

Basic Concepts in Environmental Geochemistry of Sulfidic Mine-Waste Management

Bernhard Dold
Institute of Applied Economic Geology (GEA), University of Concepción
Chile

1. Introduction

1.1 Mining and the environment

As minerals, which are essential to industrial economies, are presently not in short supply, nor do they seem to be for the next several generations, mining and mineral processing can no longer be presumed to be the best of all possible uses for land; it must compete with compelling demands for alternative uses. Environmental protection and rehabilitation are fast becoming high priorities throughout the world, no longer confined to industrialized countries. Environmental regulations in the developed countries are one of the main reasons for the departure of metal mining companies to less developed nations in the last few decades. Low labor costs, exploration potential, and lax or no existing environmental policies, reinforced this process (Hodges, 1995). While industrialized countries started to formulate environmental reports and to implement environmental framework laws in the 1970's (e.g., USA, Central Europe, Japan), developing countries started this process only recently in the 1990's (e.g., Chile, Peru, Korea, Nigeria), as reported in (Jänicke & Weidner, 1997). Increasing world population together with economic growth in emerging countries (e.g. China, India, Brasil etc.) is increasing constantly the demand for metals and minerals in the near future and the associated environmental assessment. Although the world economic crisis stopped this trend for one year, the newly increasing metal prices on the world marked confirm that this trend will go on.

1.2 Mining and extraction processes

Once the exploration of an ore body is successful, exploitation begins. The extraction of the ore can take place in an open pit or underground. The ore is then transported to stockpiles or directly to the milling process, where crushing and grinding decrease the grain size for the benefaction process. The ore grinding must be optimized with respect to the leaching, roast-leaching, or possible benefaction circuits as tabling, flotation, high intensity magnetic separation, heavy media, and others. Liberation of the mineral by the process is governed by the grain size and the mineral complexity of the ore (Ritcey, 1989).

Flotation circuits are systems of cells and auxiliary equipment arranged to yield optimal results (about 80-90 % recovery) from an ore in creating a concentrate following grinding and reagent treatment (Fig. 1). Froth flotation involves the aggregation of air bubbles and mineral particles in an aqueous medium with subsequent levitation of the aggregates to the surface and transfer to a froth phase. Whether or not bubble attachment and aggregation occur is determined by the degree to which a particle's surface is wetted by water. When a solid surface shows little affinity for water, the surface is said to be hydrophobic, and an air bubble will attach to the surface. Coal and molybdenite are the most important naturally hydrophobic minerals. For sulfides except molybdenite, and possibly stibnite, as well as non-sulfide minerals, the surface condition required for flotation is obtained by specific reagents called collectors. Furthermore, complex ores require a complex combination of conditioning, collecting, and depressing necessary for optimal mineral extraction (Weiss, 1985). After the extraction of the economically interesting minerals by flotation, the residual material (in copper mines typically 95-99% of the treated material) is transported in the form of a suspension to tailings impoundments for final deposition. In practice the recovery of sulfide minerals is less than 100%, and pyrite flotation is generally suppressed by lime addition, so the tailings resulting from sulfide ores contain certain percentage of sulfides, mostly pyrite.

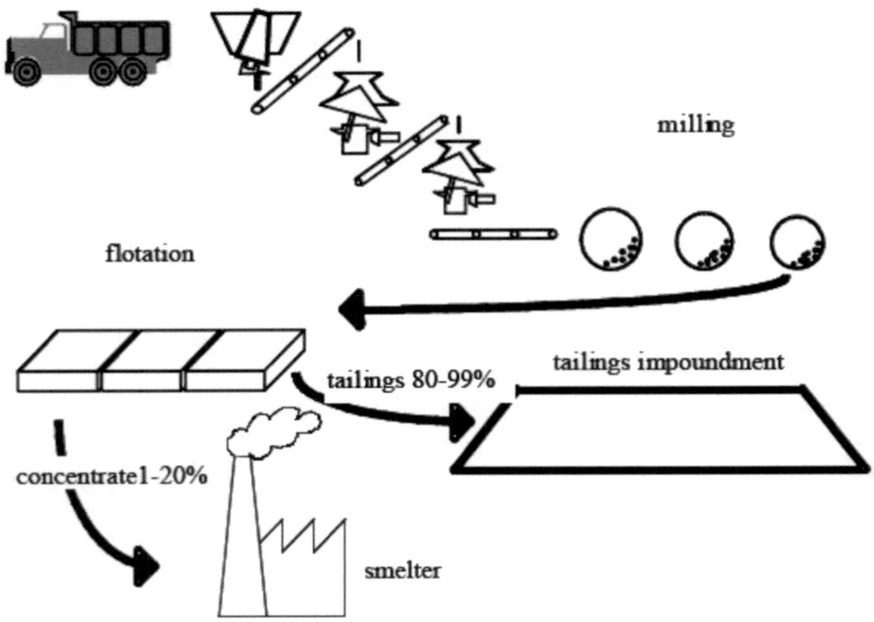

Fig. 1. The ore flow-path from the mine, through crushing and milling, to flotation. The concentrate is refined in the smelter, while the main part of the treated material will be deposited on the tailings impoundments. Tailings and concentrate proportions may differ from the given percentages.

2. Waste-rock dumps and tailings impoundment design and deposition techniques

The most economic and most commonly used tailings deposition technique is wet deposition nearby the mine site. In this technique, a water-sediment slurry is pumped in nearby topographic depressions, lakes or drainage basins. An also often economic practice was sea deposition, when the mine was located close to the ocean, with sometime hazardous effects on the environment (Dold, 2006). Other techniques are a semi-dry sub-aerial method, thickened discharge, and deep-water disposal (Ritcey, 1989).

In countries with pronounced topography (e.g., Chile), most of the tailings impoundments are designed as valley dam impoundments. This type of design is provided by placing an embankment across the valley at the head end of drainage. The most common dam construction methods are the downstream and the upstream methods. In the wet deposition method, the tailings slurry is thickened to 35-40% solids and discharged by either point or line discharge. Often a discharge point is moved periodically. As a result of a periodical move of the discharge point and gravimetrical grain size separation occurring in the tailings, a general trend of coarser to finer grain size from the tailings discharge point to the pond can be observed. Additionally, inhomogeneous layering of fine sand with silt and clay horizons makes the hydrological situation in the tailings material very complex. In general, it must be assumed that the coarser horizons are responsible for permeability, and that they behave as connected aquifers. This should be taken into account in sampling and the calculation of permeability coefficients. As the horizons have thickness in the range of centimeters to decimeters, frequently, bulk samples lead to too low permeability coefficients.

In case of Chile, the combination of valley dam impoundments with high potential energy, the extremely high seismic activity, and the fact that this country is very rich in mining resources makes the stability of the tailings dam construction the largest apparent problem. In addition, as the tailings material is generally fine-grain sized material, it retains well moisture also in arid climates, and the process of liquefaction during seismic activity is a very important issue (Byrne & Beaty, 1997). In contrast, the geochemical instability of tailings and waste rocks has received only recently in the last decade an increasing interest for research.

3. Secondary processes in sulfide mine waste - a review

3.1 Terminology

For the description of the mine waste mineralogy the classification proposed by (Jambor, 1994) is used. The term "primary" is used to designate the complete ore mineralogy, i.e. "hypogene" and "supergene". "Secondary" minerals are minerals formed within the tailings impoundment as the products of weathering processes. "Tertiary" minerals form after the sample has been removed from the tailings environment.

3.2 Sulfide oxidation in mine waste

The problem of sulfide oxidation and the associated acid mine drainage (AMD), or more generally acid rock drainage (ARD), as well as the solution and precipitation processes of metals and minerals, has been a major focus of investigation over the last 50 years (Blowes et al., 1991; Jambor, 1994; Moses et al., 1987; Nordstrom, 1982; Sato, 1960). There has been less interest in the mineralogical and geochemical interactions taking place within the tailings and waste dumps itself (Dold & Fontboté, 2001; Jambor, 1994), yet this is an essential aspect to understand the parameters controlling acid mine drainage formation and to develop effective prevention methods. The primary mineralogical composition has a strong influence on the oxidation processes. This has been best illustrated by (Huminicki & Rimstidt, 2009;

Rimstidt et al., 1994; Rimstidt & Vaughan, 2003) showing that reaction rates display significant differences depending on the sulfides being oxidized by Fe(III) and the potential Fe(III) hydroxide coating. Kinetic-type weathering experiments indicate the importance of trace element composition in the stability of individual sulfides. Where different sulfides are in contact with each other, electrochemical processes are likely to occur and influence the reactivity of sulfides (Kwong, 1993).

Most mining operations are surrounded by piles or impoundments containing pulverized material or waste from the benefaction process, which are known as tailings, waste rock dumps or stockpiles. Waste rock dumps contain generally material with low ore grade, which is mined but not milled (Run of mine). These materials can contain still large concentrations of sulfide minerals, which may undergo oxidation, producing a major source of metal and acid contamination (Dold et al., 2009). The complex microbiological, hydrological, mineralogical, and geochemical post-depositional processes and their coupled interaction in mine waste environment are not yet completely understood. In the following section the focus is on the acid producing sulfide minerals, other acid producing processes, and the acid neutralizing processes, as well as on the controlling factors for element mobility.

3.2.1 Acid producing sulfide minerals

3.2.1.1 Pyrite (FeS_2)

The most common sulfide mineral is pyrite (FeS_2). Oxidation of pyrite takes place in several steps including the formation of the metastable secondary products ferrihydrite ($5Fe_2O_3 \cdot 9H_2O$), schwertmannite (between $Fe_8O_8(OH)_6SO_4$ and $Fe_{16}O_{16}(OH)_{10}(SO_4)_3$), and goethite ($FeO(OH)$), as well the more stable secondary jarosite ($KFe_3(SO_4)_2(OH)_6$), and hematite (Fe_2O_3) depending on the geochemical conditions (Bigham et al., 1996; Cornell & Schwertmann, 2003; Dutrizac et al., 2000; Jambor, 1994; Jerz & Rimstidt, 2004; Nordstrom, 1982; Rimstidt & Vaughan, 2003). Oxidation of pyrite may be considered to take place in three major steps: (1) oxidation of sulfur (equation 1); (2) oxidation of ferrous iron (equation 2); and (3) hydrolysis and precipitation of ferric complexes and minerals (equation 4). The kinetics of each reaction are different and depend on the conditions prevailing in the tailings.

$$FeS_2 + {}^{7}/_{2}O_2 + H_2O \rightarrow Fe^{2+} + 2SO_4^{2-} + 2H^+ \qquad (1)$$

$$Fe^{2+} + {}^{1}/_{4}O_2 + H^+ \Leftrightarrow Fe^{3+} + {}^{1}/_{2}H_2O \qquad (2)$$

reaction rates strongly increased by

microbial activity (e.g., *Acidithiobacillus spp. or Leptospirillum spp.*)

$$FeS_2 + 14\ Fe^{3+} + 8H_2O \rightarrow 15\ Fe^{2+} + 2SO_4^{2-} + 16H^+ \qquad (3)$$

Equation (1) describes the initial step of pyrite oxidation in the presence of atmospheric oxygen. Once ferric iron is produced by oxidation of ferrous iron, oxidation, which may be, especially at low pH conditions, strongly accelerated by microbiological activity (equation 2), ferric iron will be the primary oxidant (equation 3) of pyrite (Ehrlich, 1996; Moses et al., 1987; Nordstrom et al., 1979). Under abiotic conditions the rate of oxidation of pyrite by ferric iron is controlled by the rate of oxidation of ferrous iron, which decreases rapidly with decreasing pH. Below about pH 3 the oxidation of pyrite by ferric iron is about ten to a hundred times faster than by oxygen (Ritchie, 1994).

It has been known for 50 years that microorganisms like *Acidithiobacillus ferrooxidans* or *Leptospirillum ferrooxidans* obtain energy by oxidizing Fe^{2+} to Fe^{3+} from sulfides by catalyzing this reaction (Bryner et al., 1967) and this may increase the rate of reaction (2) up to the factor of about 100 over abiotic oxidation (Singer & Stumm, 1970). More recent results show that a complex microorganism community is responsible for sulfide oxidation (Ehrlich, 1996; Johnson, 1998; Johnson, 1999; Johnson & Hallberg, 2003a; Norris & Johnson, 1998). Nordstrom & Southam (1997) state that the initiating step of pyrite oxidation does not require an elaborated sequence of different geochemical reactions that dominate at different pH ranges. *Acidithiobacillus* spp. form nanoenvironments to grow on sulfide mineral surfaces. These nanoenvironments can develop thin layers of acidic water that do not affect the bulk pH of the water chemistry. With progressive oxidation, the nanoenvironments may change to microenvironments. Evidence of acidic microenvironments in the presence of near neutral pH for the bulk water can be inferred from the presence of jarosite (this mineral forms at pH around 2) in certain soil horizons where the current water pH is neutral (Carson et al., 1982). Barker et al. (1998) observed microbial colonization of biotite and measured pH in microenvironments in the surroundings of living microcolonies. The solution pH decreased from near neutral at the mineral surface to 3-4 around microcolonies living within confined spaces at interior colonized cleavage planes.

When mine water, rich in ferrous and ferric iron, reaches the surface it will fully oxidize, hydrolyze and may precipitate to ferrihydrite (fh), schwertmannite (sh), goethite (gt), or jarosite (jt) depending on pH-Eh conditions, and availability of key elements such as potassium and sulfur (Fig. 2). Jarosite, schwertmannite and ferrihydrite are meta-stable with respect to goethite (Bigham et al., 1996).

The hydrolysis and precipitation of iron hydroxides (and to a lesser degree, jarosite) will produce most of the acid in this process. If pH is less than about 2, ferric hydrolysis products like $Fe(OH)_3$ are not stable and Fe^{3+} remains in solution:

$$Fe^{3+} + 3\,H_2O \Leftrightarrow Fe(OH)_3(s) + 3H^+ \tag{4}$$

Note that the net reaction of complete oxidation of pyrite, hydrolysis of Fe^{3+} and precipitation of iron hydroxide (sum of reactions 1, 2 and 4) produces 4 moles of H^+ per mole of pyrite (in case of $Fe(OH)_3$ formation, see equation 5, i.e., pyrite oxidation is the most efficient producer of acid among the common sulfide minerals (net reaction 5; Table 2). Nevertheless, it is important to be aware that the hydrolysis of $Fe(OH)_3$ is the main acid producer ($^3/_4$ of moles of H^+ per mol pyrite).

$$FeS_2 + {}^{15}/_4O_2 + {}^7/_2H_2O \rightarrow Fe(OH)_3 + 2SO_4^{2-} + 4H^+ \tag{5}$$

The precipitation of these secondary Fe(III) hydroxides can form coatings (Huminicki & Rimstidt, 2009) and cemented layers (Blowes et al., 1991; Dold et al., 2009), which can decrease the oxidation rates. (Evangelou & Zhang, 1995) reported increased oxidation rates of pyrite by addition of HCO_3^- due to the formation of pyrite surface Fe(II)-CO_3 complexes. This means that the frequent applied limestone treatment for mine waste management must be discussed critically, because if it is able to neutralize the acid produced, it can also increase the kinetic of pyrite oxidation.

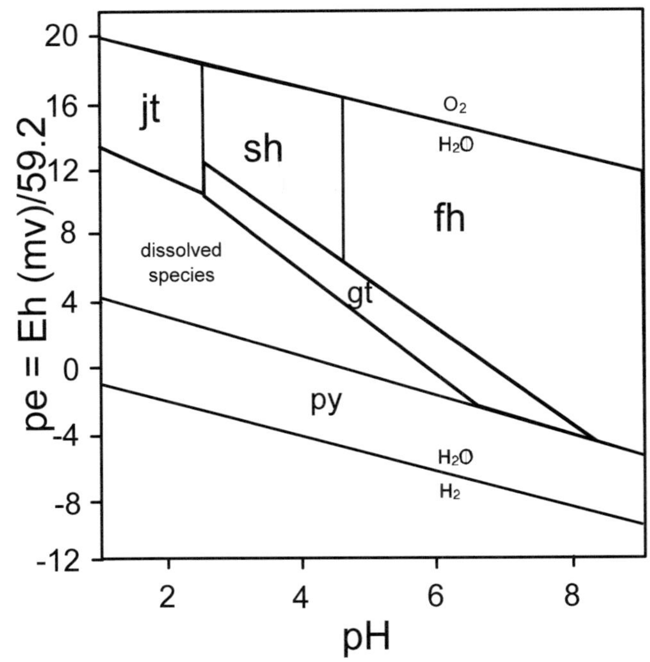

Fig. 2. pe-pH diagram for Fe-S-K-O-H system at 25°C where pe = Eh(mV)/59.2; total log activities of Fe^{2+} = -3.47; Fe^{3+} = 3.36 or −2.27; SO_4^{2-} = -2.32; K^+ = -3.78; log Kso values for solid phases for Gt = goethite, Jt = K-jarosite, Fh = ferrihydrite, Sh = schwertmannite are 1.40, -12.51, 4.5 and 18.0, respectively. Py = pyrite. Mean composition of the schwertmannites used for the development of this pe-pH diagram was $Fe_8O_8 (OH)_{4.8} (SO_4)_{1.6}$. After Bigham et al. (1996). Note that these stability fields have to be interpreted as indicative, as the thermodinamic data publihed from schwertmannite and ferrihydrite show high veriability (Majzlan et al., 2004).

Marcasite, the orthorhombic dimorph of pyrite, may also abundantly occur as a primary mineral in sulfidic tailings, mainly from sediment-hosted base metal deposits and as primary alteration product of pyrrhotite. Marcasite has the same formula as pyrite and leads to the same amount of acid production via oxidation. Direct observation (e.g. the Zn-Pb deposit of Reocín, Spain) suggest that its oxidation kinetics should be faster than that of pyrite, perhaps as a result of its typical twinned morphology and finer grain size, which offers more surface to oxidation than pyrite. Additional, (Jambor, 1994) reports the presence of marcasite as a secondary alteration product of pyrrhotite in zonal patterns in oxidizing mine tailings.

3.2.1.2 Pyrrhotite(Fe(1-x)S)

Wastes from sulfide ores often contain pyrrhotite associated with pyrite. The general formula of pyrrhotite is $Fe_{(1-x)}S$, where x can vary from 0.125 (Fe_7S_8) to 0.0 (FeS, troilite). The oxidation rates and weathering products of pyrite are well known, but only a few investigations have focused the oxidation of pyrrhotite. (Nicholson & Scharer, 1994) propose for the oxidation of pyrrhotite the following equation:

$$Fe_{(1-x)}S + (2-x/2)O_2 + xH_2O \rightarrow (1-x)Fe^{2+} + SO_4^{2-} + 2xH^+ \qquad (6)$$

in which the stoichiometry of the pyrrhotite affects the relative production of acid. At one extreme, if x = 0 and the formula is FeS, no H^+ will be produced in the oxidation reaction; at the other extreme, the maximum amount of acid will be produced by the iron-deficient Fe_7S_8 phase. The main part of acid is produced by the oxidation of pyrite (equation 1) and the subsequent hydrolysis of ferric hydroxides (equation 4). In conclusion, the role of pyrrhotite in the acidifying process is similar to that of pyrite, but it is very important at early weathering stages because its oxidation rate is 20 to 100 times higher than that of pyrite in atmospheric concentrations of O_2 and at 22°C (Nicholson & Scharer, 1994).
The oxidation of pyrrhotite can also involve the formation of elemental sulfur (Ahonen & Tuovinen, 1994), marcasite (Jambor, 1994), or the formation of pyrite. They are acid-consuming reactions.

3.2.1.3 Chalcopyrite ($CuFeS_2$)

Complete oxidation of chalcopyrite may be written as:

$$2CuFeS_2 + 4O_2 \rightarrow 2Cu^{2+} + Fe^{2+} + SO_4^{2-} \quad (7)$$

without acid production. Nevertheless, the combination of ferrous iron oxidation and ferrihydrate hydrolysis will be again the main acid producing process.

$$2CuFeS_2 + {}^{17}/{}_{2}O_2 + 5\,H_2O \rightarrow 2Cu^{2+} + 2Fe(OH)_3 + 4\,SO_4^{2-} + 4H^+ \quad (8)$$

Chalcopyrite, together with molybdenite, is known as one of the most resistant sulfides to oxidation (Plumlee, 1999). (Rimstidt et al., 1994) reported that the oxidation rate of chalcopyrite increases with increasing ferric iron concentration, but with an oxidation rate of 1-2 orders of magnitude less than pyrite.

3.2.1.4 Arsenopyrite (FeAsS)

Arsenopyrite (FeAsS) may be oxidized by the following reaction path (Mok & Wai, 1994):

$$4FeAsS + 13O_2 + 6H_2O \rightarrow 4Fe^{2+} + 4\,SO_4^{2-} + 4H_2AsO_4^- + 4H^+ \quad (9)$$

Combined with ferrous iron oxidation and ferrihydrate precipitation, the overall arsenopyrite oxidation reaction can be written as follow:

$$FeAsS + {}^{7}/{}_{2}O_2 + 6H_2O \rightarrow Fe(OH)_3 + SO_4^{2-} - + H_2\,AsO_4^- + 3H^+ \quad (10)$$

If ferric iron is the oxidant, the oxidation rate of arsenopyrite is similar to the oxidation rate of pyrite. If it is oxygen, the oxidation rate of arsenopyrite is somewhat lower than that of pyrite (Mok & Wai, 1994).

3.3.2 Non-acid producing sulfide minerals

3.3.2.1 Sphalerite (ZnS) and galena (PbS)

Sphalerite and galena are the most important base metal bearing minerals. Though Zn is toxic only at very high concentrations, sphalerite may contain environmentally dangerous amounts of Cd and Thallium (Tl). In addition, Fe may significally substitute for Zn, in cases up to 15 mole %, in sphalerite (Baumgartner et al., 2008). If iron substitutes for zinc, sphalerite will be an acid generator in a similar way as pyrrhotite, due to hydrolysis of ferric phases. Galena is the main source of Pb contamination in mine areas. The common result of

Mineral	Mole H^+/Mole Mineral	Relative wt % H^+/Mole mineral	Reaction with Fe(III)[1)]
Pyrite	4	0.03	2.7×10^{-7}
Marcasite	4	0.03	1.5×10^{-7}
Arsenopyrite	2	0.018	1.7×10^{-6}
Chalcopyrite	2	0.011	9.6×10^{-9}
Pyrrhotite	2-0	0.022	-
Enargite	1	0.002	-

[1)] $m_{Fe^{3+}} = 10^{-3}$ and pH = 2.5 at 25□C; mol $m^{-2}s^{-1}$; - = not studied by Rimstidt et al. (1994)

Table 1. Hydrogen ions per mole produced by oxidation via O_2 of some frequent minerals in mine tailings (Plumlee, 1999) and some laboratory reaction rate data from Rimstidt et al. (1994). However, Jambor (1994) suggested the relative resistance of sulfide in oxidizing tailings environment to follow the increasing order pyrrhotite → sphalerite-galena → pyrite-arsenopyrite → chalcopyrite → magnetite.

the wet oxidation of sphalerite is a leach rich in dissolved Zn, and sulfate, with variable Cd amounts, while that of galena is secondary anglesite ($PbSO_4$) in equilibrium with a Pb^{2+} and SO_4^{2-} solution according to the following equations

$$ZnS + 2O_2 \rightarrow Zn^{2+} + SO_4^{2-} - \quad (11)$$

$$PbS + 2O_2 \rightarrow Pb^{2+} + SO_4^{2-} \quad (12)$$

Secondary anglesite coating on galena may increase the apparent resistance because anglesite has a relative low solubility and protects the sulfides from direct contact with oxidizing reagents (Jambor & Blowes, 1998). In the presence of Fe^{3+}, the oxidation of MeS (where Me = divalent metal) produces acidity according to reaction schemes where part of the oxidation capacity of the system is derived from Fe^{3+} as, e.g.

$$2MeS + 4Fe^{3+} + 3O_2 + 2H_2O \rightarrow 2Me^{2+} + 4Fe^{2+} + 2\,SO_4^{2-} + 4H^+ \quad (13)$$

Plumlee (1999) pointed out that the oxidation of sulfide minerals by aqueous ferric iron generates significantly greater quantities of acid than the oxidation by oxygen (e.g., equation 3 and 13). This is correct in the case that ferric iron is added to the system (e.g., in form of primary ferric minerals or seepage of ferric iron containing solutions). However, if we consider that the ferric iron is produced by the oxidation from ferrous iron in the system (e.g., microbiologically catalyzed pyrite oxidation), for every mole of ferric iron produced one mole of protons is consumed (equation 2). This leads to the same overall produced quantity of two protons per mole pyrite oxidized as in case of pyrite oxidation via O_2 (compare equations 1, 2, and 3). Additionally, in case of oxidation via ferric iron no acidity can be produced via the hydrolysis of the ferric phases, the main acid producing process in sulfide oxidation (see equation 4). Thus, the sulfide oxidation by ferric iron has faster kinetics and is able to oxidize sulfide minerals in the absence of oxygen, but it does not produce more acid as the oxidation via oxygen when the ferric iron is produced in the system. However, the management of ferric iron rich solutions in mine waste environments can lead to strong exportation of acid potential and has to receive therefore special attention (Dold et al., 2009).

Table 1 shows a summary of the hydrogen ions produced by some common sulfide minerals by the oxidation via O_2. There are important differences and therefore exact knowledge of the sulfide minerals contained in waste rocks or tailings is crucial for adequate acid-base accounting (ABA).

3.3.3 Secondary Fe(III) hydroxides, oxyhydroxides, and oxyhydroxide sulfates

As explained in 3.2.1, the acid production processes can be split into two parts. The first is the oxidation of sulfide minerals by oxygen and ferric iron. The second is the hydrolysis of mainly Fe(III) and subsequent precipitation of ferric oxyhydroxides or oxyhydroxide sulfates.

The process of hydrolysis of Fe(III) has been reviewed by Sylva (1972), Flynn (1984), Schneider & Schwyn (1987), Stumm & Morgan (1996), and Cornell & Schwertmann (2003). Metal ions undergo hydrolysis because coordinated water is a stronger acid than free water. This result from the effect that the metal-oxygen bond weakens the O-H bonds in a way that in aqueous systems the free water molecules behave as proton acceptors. So the hydrolysis of metal ions is the result of the deprotonation of the coordinated water molecules (Sylva, 1972).

The hydrolysis, i.e. deprotonation, starts with the hexa-aquo ion (Fe(III) is hydrated by 6 water molecules), except at very low pH, where Fe^{3+} is stable. Initially, low molecular weight species such as $Fe(OH)_2^{+}$ and $Fe(OH)_2^{1+}$ form rapidly. The following dissolved species will be stable depending on pH (see also Fig. 3 and Table 2): Fe^{3+}, $Fe(OH)^{2+}$, $Fe(OH)_2^{+}$, $Fe(OH)_3^{0}$ (aq), $Fe(OH)_4^{-}$, $Fe_2(OH)_2^{4+}$, and $Fe_3(OH)_4^{5+}$ (Stumm and Morgan, 1996). By "aging" increasingly complex polymers, Fe(III) oxides, oxyhydroxides, or oxyhydroxide sulfates like ferrihydrite, schwertmannite, goethite, jarosite are formed (Fig. 2; Bigham et al., 1996). Depending on the secondary phases precipitated, considerably different amounts of protons are produced (Table 3).

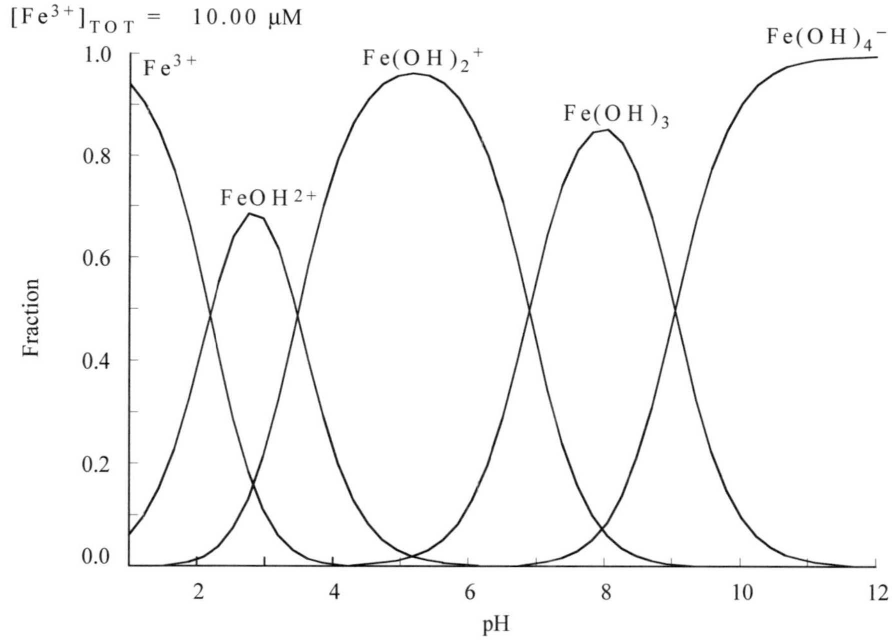

Fig. 3. Mole fraction of total dissolved Fe(III) present as Fe^{3+} and Fe(III)-OH complexes as a function of pH in pure water at 25°C (after Langmuir (1997)).

Species	Equation	Fe^{3+}	H^+	log *K* (I = 3 M)
Fe^{3+}		1	0	0
$Fe(OH)^{2+}$	$Fe^{3+} + H_2O \Leftrightarrow Fe(OH)^{2+} + H^+$	1	-1	-3.05
$Fe(OH)_2^+$	$Fe^{3+} + 2H_2O \Leftrightarrow Fe(OH)_2^+ + 2H^+$	1	-2	-6.31
$Fe(OH)_{3\,(aq)}$	$Fe^{3+} + 3H_2O \Leftrightarrow Fe(OH)_{3\,(aq)} + 3H^+$	1	-3	-13.8
$Fe(OH)_4^-$	$Fe^{3+} + 4H_2O \Leftrightarrow Fe(OH)_4^- + 4H^+$	1	-4	-22.7
$Fe_2(OH)_2^{4+}$	$2Fe^{3+} + 2H_2O \Leftrightarrow Fe_2(OH)_2^{4+} + 2H^+$	2	-2	-2.91
$Fe_3(OH)_4^{5+}$	$3Fe^{3+} + 4H_2O \Leftrightarrow Fe_3(OH)_4^{5+} + 4H^+$	3	-4	-5.77

Table 2. Hydrolysis reactions of Fe(III) species and the associated protons produced (after Stumm & Morgan, 1996).

Phase	Equation	moles H^+/mole Fe^{3+}hydrolyzed
amp. $Fe(OH)_{3(s)}$	$Fe^{3+} + 3H2O \Leftrightarrow Fe(OH)_{3(s)} + 3H^+$	3
ferrihydrite	$10\ Fe^{3+} + 60H2O \Leftrightarrow 5Fe2O3{\cdot}9H2O + 30H^+$	3
goethite	$Fe^{3+} + 2H2O \Leftrightarrow FeO(OH) + 3H^+$	3
hematite	$2Fe^{3+} + 3H2O \Leftrightarrow Fe2O3 + 6H^+$	3
schwertmannite	$8Fe^{3+} + SO_4^{2-} + 14H2O \Leftrightarrow Fe8O8\ (OH)\ 6SO4 + 22H^+$ $16Fe^{3+} + 3SO_4^{2-} + 26H2O \Leftrightarrow Fe16O16(OH)10(SO4)3 + 42H^+$	2.75 2.625
jarosite	$3Fe^{3+} + K + 2SO_4^{2-} + 6H2O \Leftrightarrow KFe3(SO4)2\ (OH)6 + 6H^+$	2

Table 3. Protons produced by the hydrolysis of the different secondary Fe(III) phases.

3.3.4 Dissolution of sulfate minerals

3.3.4.1 Iron sulfate minerals

Iron sulfate minerals are the most common secondary minerals found in the oxidizing environment of the mine waste, due to the wide distribution of pyrite and pyrrhotite as a iron and sulfur source. They are also common in the oxidized portions of weathering zones from ore deposits (e.g., gossans). They may be composed of Fe(II), Fe(II) + Fe(III), or only Fe(III). Examples are melanterite, roemerite, and coquimbite, respectively (Alpers et al., 1994; Nordstrom & Alpers, 1999). In general they have a high solubility, whereas supergene jarosite shows relatively low dissolution kinetics. Baron and Palmer, (1996) conducted a series of dissolution experiments with jarosite under 4-35°C and at pH values between 1.5 and 3. Equilibrium was established in the experiment after approximately 3 to 4 months. It is important to notice that this mineral group is meta-stable in respect to more stable iron hydroxides and oxides and may liberate acidity by this transformation (see equation 14-16). Transformation:

$$\text{jarosite} \Rightarrow \text{goethite: } KFe_3(SO_4)_2(OH)_6 \Longrightarrow 3FeO(OH) + K^+ + 2SO_4^{2-} + 3H^+ \quad (14)$$

$$\text{schwertmannite} \Rightarrow \text{goethite: } Fe_{16}O_{16}(OH)_{10}(SO_4)_3 + 6H_2O \Rightarrow 16FeO(OH) + 3SO_4^{2-} + 6H^+ \quad (15)$$

$$\text{schwertmannite} \Rightarrow \text{goethite: } Fe_8O_8(OH)_6(SO_4) + 2H_2O \Rightarrow 8FeO(OH) + SO_4^{2-} + 2H^+ \quad (16)$$

3.3.4.2 Water-soluble sulfates (efflorescent salts)

An extended group of highly water-soluble sulfates and chlorides, as for example gypsum or chalcanthite ($CuSO_4 \cdot 5H_2O$) or eriochalcite ($CuCl_2 \cdot 2H_2O$), formed under oxidizing conditions and high evaporation rates, can release significant amounts of metals and acid with rain. This mineral group is an important factor leading to seasonal fluctuations in contamination levels of ground and surface waters, especially in semi-arid and arid climates (Alpers et al., 1994; Dold, 2006; Dold & Fontboté, 2001). Additionally, this group of water-soluble minerals expose heavy metals to aeolian transport in arid climates in a highly bioavailable form and have to receive therefore special attention in health risk evaluations (Dold, 2006).

3.4 Neutralization processes

The acid produced in the processes presented above may result normally in pH in the range of 1.5 - 4 in the mine waste environment. Exceptionally, pH can reach even negative values (Nordstrom et al., 2000). This acidity together with Fe(III), are able to dissolve minerals and mobilize elements in the tailings (Dold & Fontboté, 2001). In their pathway, the acid produced and the elements mobilized react with acid-neutralizing minerals such as carbonates or silicates. Acid-neutralizing reactions result in an increase in the pore-water pH. This increase in pH is frequently accompanied by precipitation of metal-bearing oxyhydroxide and oxyhydroxide sulfate minerals that remove dissolved metals from the water migrating within the tailings pore space. These secondary minerals act in a certain pH range as buffers so that a sequence of pH buffering reactions can be observed in the tailings environment (Blowes & Ptacek, 1994).

3.4.1 Carbonates

Dissolution of carbonates releases alkaline earth and metal cations, including Ca, Mg, Fe, and Mn. These cations participate in the formation of secondary solids, including simple hydroxide solids, which in some cases can later dissolve and contribute to acid neutralization. The ability of carbonates to neutralize acid by fast reaction makes them an important part of the mineralogical assemblage for ARD prediction and prevention.

3.4.1.1 Calcite ($CaCO_3$)

Calcite is the most common carbonate mineral and the fastest reacting. Its solubility depends on the proton concentration as shown in the following equations:

$$CaCO_3 + H+ \Leftrightarrow Ca^{2+} + HCO_3^- \quad (17)$$

This reaction will buffer at near neutral pH (6.5 - 7), while in more acidic environments the following equation can be written:

$$CaCO_3 + 2H^+ \Leftrightarrow Ca^{2+} + H_2CO_3^0 \quad (18)$$

Carbonate speciation is pH dependent (Fig. 4) and the dissolution of calcite increases the amount of carbonate in solution, increasing therefore the neutralization potential of the solution. It is important to mention that calcite buffers the pH to neutral values and at pH 7 HCO_3^- is the dominant specie (Fig. 4). Thus, for the neutralization of 1 mol H^+ 1 mol calcite is necessary as shown in equation (17). At pH below pH 6.3 carbonate can neutralize two moles of protons, as $H_2CO_3^0$ is the dominant specie (Equation 18). When neutralization proceeds and pH increases, calcite may precipitate as a secondary mineral.

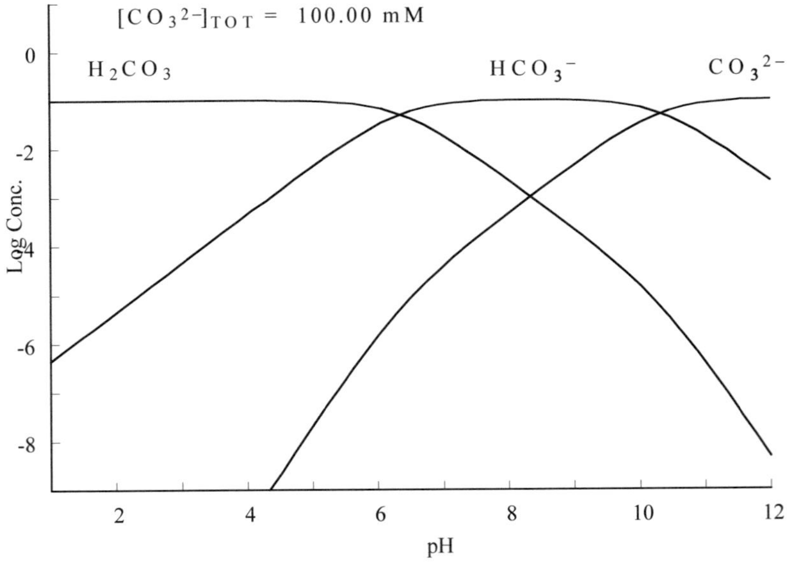

Fig. 4. Distribution of solute species in the aqueous carbonate system. Log concentration vs. pH for 100 mM CO_3^{2-}.

3.4.1.2 Siderite ($FeCO_3$)

Siderite may act under certain conditions as a neutralizer, and under other conditions as an acid producer. The following dissolution reaction may take place and buffer the system at pH around 5:

$$FeCO_3 + 2H+ \Leftrightarrow Fe^{2+} + H_2CO_3^- \tag{19}$$

Combined with ferrous iron oxidation and ferrihydrate precipitation, the overall siderite oxidation reaction is:

$$4FeCO_3 + O_2 + 10H_2O \Leftrightarrow 4\ HCO_3^- + 4Fe(OH)_3 + 4H^+ \tag{20}$$

This indicates that under elevated pH, where bicarbonate is stable, the total reaction from dissolution of siderite to the precipitation of ferrihydrate gives a net acid production of one mole hydrogen ion per mol siderite dissolved. However, under more acidic conditions, where carbonic acid is stable, there will be no net acid production. However, it has to be considered that ferrous iron is liberated, which can be mobilized under neutral pH condition and, when outcropping downstream under oxidizing conditions it will oxidize and hydrolyze under the liberation of protons, producing AMD far from source.

If ferrous iron is present in a solution containing bicarbonate, the formation of siderite may occur by the following reaction:

$$Fe^{2+} + HCO_3^- \Leftrightarrow FeCO_3 + H^+ \tag{21}$$

This reaction buffers the pH at around 5 - 5.5. This reaction path may be an alternative to ferrous to ferric iron oxidation and the consecutive precipitation of Fe(III)oxyhydroxides. Geochemical studies by several authors have shown that in tailings impoundmends and AMD affected aquifers the waters are frequently close to saturation or even oversaturated with respect to siderite (Blowes et al., 1994; Blowes et al., 1991; Morin & Cherry, 1986).

3.4.2 Lime ($Ca(OH)_2$)

Lime is a common agent in the flotation circuit to depress the flotation of pyrite by the increase of the pH to 10.5.

$$Ca(OH)_3 + 2H^+ \Leftrightarrow Ca^{2+} + 2H_3O \quad (22)$$

$$Ca(OH)_3 \Leftrightarrow Ca^{2+} + 2OH^- \quad (23)$$

Lime is used in hydrometallurgical processes due to its high solubility in water and fast neutralization reaction. It may be washed out from tailings impoundments. In this case, the role in neutralizing the acid produced in the tailings may be minor. Therefore, for pH buffering in mine wastes (tailings and waste dumps) carbonates should be preferred.

3.4.3 Metal hydroxides dissolution

As a result of neutralization and pH increase, the precipitation of metal hydroxides or hydroxide sulfates is favored, as gibbsite, amorphous $Al(OH)_3$, amorphous $Fe(OH)_3$, ferrihydrite, goethite, or schwertmannite. Some of the reactions can be described as follows:

$$Al(OH)_3 + 3H^+ \Leftrightarrow Al^{3+} + 3H_2O \text{ this reaction buffers the pH to 4.0 - 4.3} \quad (24)$$

$$Fe(OH)_{3(s)} + 3H^+ \Leftrightarrow Fe^{3+} + 3H_2O \text{ this reaction buffers the pH to below 3.5} \quad (25)$$

$$FeO(OH)_{(s)} + 3H^+ \Leftrightarrow Fe^{3+} + 3H_2O \text{ this reaction buffers the pH to below 3.5} \quad (26)$$

These reactions complement an ideal neutralization sequence, which starts with calcite (pH 6.5 - 7.5) followed by siderite (pH 5.0 - 5.5). When all carbonates are consumed the next buffer is gibbsite (pH 4.0 - 4.3), followed by Fe(III)hydroxides as goethite (pH below 3.5). In the oxidation zone itself the pH range commonly between pH 2 and 3 depending on the amount of goethite, schwertmannite and jarosite (Dold & Fontboté, 2001).

3.4.4 Silicates

Dissolution of most aluminosilicate minerals also consumes H^+ ions and contribute base cations (Ca, Mg, Fe(II)), alkali elements (Na, K) and dissolved Si and Al to the tailings pore water (Blowes & Ptacek, 1994). Though, dissolution of aluminosilicate minerals is slower than of metal hydroxides and much slower than that of carbonates (Table 4). Feldspar weathering is mainly controlled by pH, silica, Na, K, and Ca concentrations. One possible reactions path is:

K-feldspar -> kaolinite -> gibbsite

Reactions 27 and 28 illustrate this path.

$$\underset{\text{K-feldspar}}{2KAlSiO_3O_8} + 9H_2O + 2H^+ \rightarrow \underset{\text{kaolinite}}{Al_2Si_2O_5(OH)_4} + 2K^+ + 4H_4SiO_4 \quad (27)$$

In this reaction, K and Si enter the solution, while protons are consumed. The solubility of feldspar increases when proton activity increases or the removal of K or Si is fast, e.g. by secondary mineral formation. Secondary kaolinite may dissolve to form gibbsite, a reaction that does not neutralize acid:

$$AlSi_2O_5(OH)_4 + 5H_2O \rightarrow 2Al(OH)_3 + 2H_4SiO_4 \quad (28)$$

kaolinite gibbsite

Higher pH and K concentrations can lead theoretically to muscovite formation instead of kaolinite as secondary mineral of feldspar weathering:

$$3KAlSiO_3O_8 + 12H_2O + 2H^+ \rightarrow KAl_2(AlSi_3O_{10})(OH)_2 + 2K^+ + 6H_4SiO_4 \quad (29)$$

K-feldspar muscovite

and muscovite reacts forming gibbsite:

$$KAl_2(AlSi_3O_{10})(OH)_2 + 9H_2O + H^+ \rightarrow 2Al(OH)_3 + 3H_4SiO_4 + K^+ \quad (30)$$

muscovite gibbsite

However, these reactions have to be seen as a strong simplification of the very complex clay mineral group and their formation processes. In the case of plagioclase, the weathering path is similar to that of K-feldspar and is accompanied by the release of sodium and/or calcium. Under low pH conditions plagioclase will react to form kaolinite, while under higher pH smectite will be formed.

Group	Typical Minerals	Relative Reactivity (pH5)
1. Dissolving	calcite, dolomite, magnesite, aragonite, brucite,	1.0
2. Fast weathering	anorthite, olivine, garnet, diopside, wollastonite, jadeite, nepheline, leucite, spodumene	0.6
3.Intermediate weathering	enstatite, augite, hornblende, tremolite, actinolite, biotite, chlorite, serpentine, talc, epidote, zoisite, hedenbergite, glaucophane, anthophyllite	0.4
4. Slow weathering	plagioclase (Ab_{100}-Ab_{30}), kaolinite, vermiculite, montmorillonite, gibbsite	0.02
5. Very slow weathering	K-feldspar, muscovite	0.01
6. Inert	quartz, rutile, zircon	0.004

Table 4. Relative reactivity in acid-neutralization capacity of minerals (after Jambor & Blowes, 1998).

Nesbitt & Jambor (1998) have shown the fundamental role of mafic minerals in neutralization of the Waite-Amulet tailings. As in the weathering of feldspar, the weathering of felsic minerals leads to the formation of clay minerals. Muscovite, pyroxene, and amphibole alter to chlorite. By decreasing pH, chlorite alters to sericite, kaolinite or Mg-montmorillonite. The products of biotite alteration are hydrobiotite, a regularly interstratified biotite-vermiculite phase, vermiculite, and kaolinite (Acker & Bricker, 1992; Malmström & Banwart, 1997). Direct conversion of biotite to kaolinite has also been described e.g. (Acker & Bricker, 1992). The alteration from biotite to vermiculite may be written as:

$$3K_2(Mg_3Fe_3)Al_2Si_6O20)(OH)_2 + 5H_2O + 12H_2CO_3 + 6H^+ + {}^3/_2\,O_2 \rightarrow$$

$$2[(Mg_3Fe_3)\,Al_3Si_5O20(OH)_2 \cdot 8\,H_2O] + 6K^+ + 3Mg^{2+} + 3Fe^{2+} + 8SiO_2 + 12HCO_3^- \quad (31)$$

Malmström & Banwart (1997) studied the pH dependence of dissolution rate and stoichiometry of biotite at 25°C. They found that the release of the interlayer K is relatively fast and becomes diffusion controlled within a few days. The release of framework ions (Mg, Al, Fe, Si) is much slower. Strömberg & Banwart (1994) suggested that, in the absence of carbonates, primary minerals, particularly biotite, provide the major sink for acidity in drainage from mine waste rock (see equation 31). However, this statement has to be taken with care, as in the neutralization via biotite also Fe(II) is liberated (equation 31) and may so produce acidity via hydrolysis.

3.5 Dissolution

Dissolution is mainly controlled by surface complexation of protons or organic ligands such as the organic acids acetate, oxalate, or citrate. Organic acids are widely used in studies of dissolution kinetics and the solubility of secondary oxides and hydroxides, especially in soil science and acid mine drainage (Cornell & Schwertmann, 2003; Dold, 2003a). For example, the dissolution rates of Fe(III)hydroxides by oxalate are increased photochemically, by higher acidity and temperature, and by the presence of Fe(II) in the system (Suter et al., 1988). Reduction also increases dissolution kinetics because Fe(II) has a greater atomic radius (0.76 Å) than Fe(III) (0.64 Å), so Fe(II) does not fit any more in the crystalline system of the ferric minerals and the detachment of the ferrous ion is facilitated (Stumm & Sulzberger, 1992).

3.6 Prediction - Acid-Base Accounting (ABA)

It has been shown in the above sections that there are minerals able of producing acid (Acid Potential - AP) and those, which are able to neutralize acid (Neutralizing Potential - NP) in mine waste. Understanding the relative influence of these two parameters can lead to an estimation of the net acid-producing potential (NAPP) or net neutralizing potential (NNP). It is important to be able to predict if or if not a geological unit has the capacity of generating acid, and is a deciding factor for further treatment strategies of the material. Depending on the complexity of the mineral assemblage, it is a major task to understand all the interactions and processes taking place and to calculate the acid-base accounting (ABA). An ABA can be seen a an intent to quantify these two relevant mineral groups (Sulfides and carbonates), in order to predict if a rock-unit will or will not produce an acid environment. A review of the used static and kinetic test procedures is given in White III et al. (1999). A simple ABA would be to measure the total sulfur and total carbon contents in a sample and assume that the total sulfur value represents the pyrite content and the total carbon the calcite content. More elaborated methods try to take into account the different sulfur-bearing phases and to record the complexity of the acid-neutralizing mineral assemblage. For example sequential extraction procedures can be used as high-resolution ABA method (Dold, 2003b).

However, there remain mayor problem in the calculation of standard ABA. Most acid-base accounting procedures (ABA) assume that H_2CO_3 (equation 18) is the dominant specie, resulting in a convertion factor for the acid potential (AP) of 31.25 based on the sulfur concentration. This is correct at lower pH (below pH 6.3) and doubles the buffer capacity of calcite (equation 18). However, if the system should be buffered at neutral pH, equation 17 has to be considered, where one mol of calcite can neutralize only one mol of protons, leading to a convertion factor for the acid potential of 62.5. The factor 31.25 leads to an overestimation of 100% of the neutralization potential at neutral pH. Additionally, the standard ABA does not consider the specific mineralogy present in the sample (e.g. acid producing Fe(III) oxyhydroxide sulfates, like jarosite and schwertmannite).

3.7 Mobility and sorption processes

The liberation of elements from minerals depends mainly on the solubility of minerals, which act as hosts of the metals or other elements. Once the element is liberated, its mobility is controlled by the complex-species stability at the existing pH, redox and other geochemical conditions, and the surface charge of the adsorbents, which is also pH dependent. The hydroxides and clay minerals are characterized by their small grain size and high surface area combined with a net surface charge; they are therefore effective adsorbents.

3.7.1 Complexation

Stumm and Morgan (1996) define complex formation as follows: Coordination or complex formation is referred to as any combination of cations with molecules or anions containing free pairs of electrons (bases). This combination can be electrostatic, covalent, or a mixture of both. The metal cation will be called the central atom, and the anions are ligands. Two types of complex species can be distinguished: the ion pairs and the complexes. Ions of opposite charge that approach within a critical distance effectively form an ion pair and are no longer electrostatically effective. In contrast, most stable entities that result from the formation of largely covalent bonds between a metal ion and an electron-donating ligand are called complexes (Parks, 1990).

3.7.2 Stability of complex species

Chemical speciation and species refer to the actual form in which a molecule or ion is present in solution. If a metal cation is liberated into solution it will be on the search for a partner. As a result, metal cations will tend to form in water aquo- or hydroxo-complexes. This process is called hydrolysis and was discussed in the example of ferric hydroxide formation (3.2.1). Fig. 3 illustrates the predominant complex species of Fe(III) depending on the pH of the oxidation state of the central atom and of the solution.
Equilibrium in hydrolysis reactions is usually established fast, as the hydrolysis species are simple. The "aging" of the solution and the associated formation of polynuclear complexes is a slower process and can be seen as the intermediate state to the solid precipitate. Hence, hydrolysis species are thermodynamically unstable or meta-stable (Stumm & Morgan, 1996). Complexes with monodenate ligants are usually less stable than those with multidenate ligands. More important is the fact that the degree of complexation decreases more strongly with dilution for monodenate complexes than for multidenate complexes (chelates).

3.7.3 Redox reactions

The stability of species depends strongly on the reduction-oxidation (redox) reaction taking place between the ions. In a similar way that acids and bases are interpreted as proton donors and proton acceptors, reductants and oxidants are defined as electron donors and electron acceptors. Because there are no free electrons in nature, every oxidation is accompanied by a reduction, or in other words, an oxidant is a substance that causes oxidation to occur while being reduced itself.

$$O_2 + 4H^+ + 4e^- \Leftrightarrow 2H_2O \text{ reduction} \quad (32)$$

$$4Fe^{2+} \Leftrightarrow 4Fe^{3+} + 4e^- \text{ oxidation} \quad (33)$$

$$O_2 + 4Fe^{2+} + 4H^+ \Leftrightarrow 4Fe^{3+} + 2H_2O \text{ redox reaction} \quad (34)$$

The combination of redox condition (expressed as Eh or pe) and pH makes it possible to predict which species are dominant under the specific geochemical conditions. These stability fields are made visible in the Eh-pH diagrams largely used in geochemistry (e.g., Fig. 3 and (Brookins, 1988)).

3.7.4 Sorption

Sorption is a general name for adsorption, absorption, and ion exchange. Sorption also includes surface precipitation and element diffusion. The sorption processes take place at the mineral-water interface and are controlled by the reactivity of surface functional groups. Surface functional groups are the surface mineral atoms that may form chemical reactions with species in solution, forming mineral-species complexes.

Whether or not a mobilized element will be adsorbed depends on the redox conditions resulting from specific speciation of the metal complexes and on the pH dependent reactivity of the surface functional groups of the absorbent. Oxides, oxyhydroxides and silicates surfaces in contact with water typically are electrically charged because of ionization of the functional groups. The magnitude and sign of this surface charge vary with the solution pH (Parks, 1990).

Adsorption and absorption processes of metal ions on iron hydroxides, clay minerals, and calcite have been well investigated in laboratory research (Dzombak & Morel, 1990). Adsorption of metal ions on Fe(III)hydroxides is a function of pH, temperature, surface area of sorbent, dissolved metal concentrations, and reaction time (Dzombak & Morel, 1990; Stumm & Morgan, 1996). Long-term studies show that metal ions may be incorporated by diffusion into the crystalline systems of secondary ferric hydroxides (Davis & Ritchie, 1986; Donnert et al., 1990). This process, where the adsorbate becomes incorporated in the crystal structure of the adsorbent, including the formation of solid solution by co-precipitation or solid-state diffusion, is referred to as absorption (Brown et al., 1995). Adsorption can be differentiated into two processes. First, the specific adsorption or chemosorption, where the sorption at the mineral-water interface may involve further reactions of some ions, results in the loss of one or more waters of hydration from the adsorbate ion and the formation of a relatively strong chemical bond between adsorbate and adsorbent (ligand exchange). The adsorbed species is refered to as an inner-sphere adsorption complex. A weaker interaction may occur between the hydrated ion and the mineral surface in which waters of hydration are retained (Fig. 5). Adsorption of this type is termed non-specific, and the adsorbed species is referred to as an outer-sphere adsorption complex (Brown et al., 1995).

The resulting adsorption is a function of the species stable at each pH and the net surface charge of the adsorbent at the relevant pH, e.g. for goethite is the "zero point of charge" (ZPC) at pH 8. In Fig. 6 adsorption of metal cations and oxyanions are shown with the net surface charge of the adsorbent, in this case goethite.

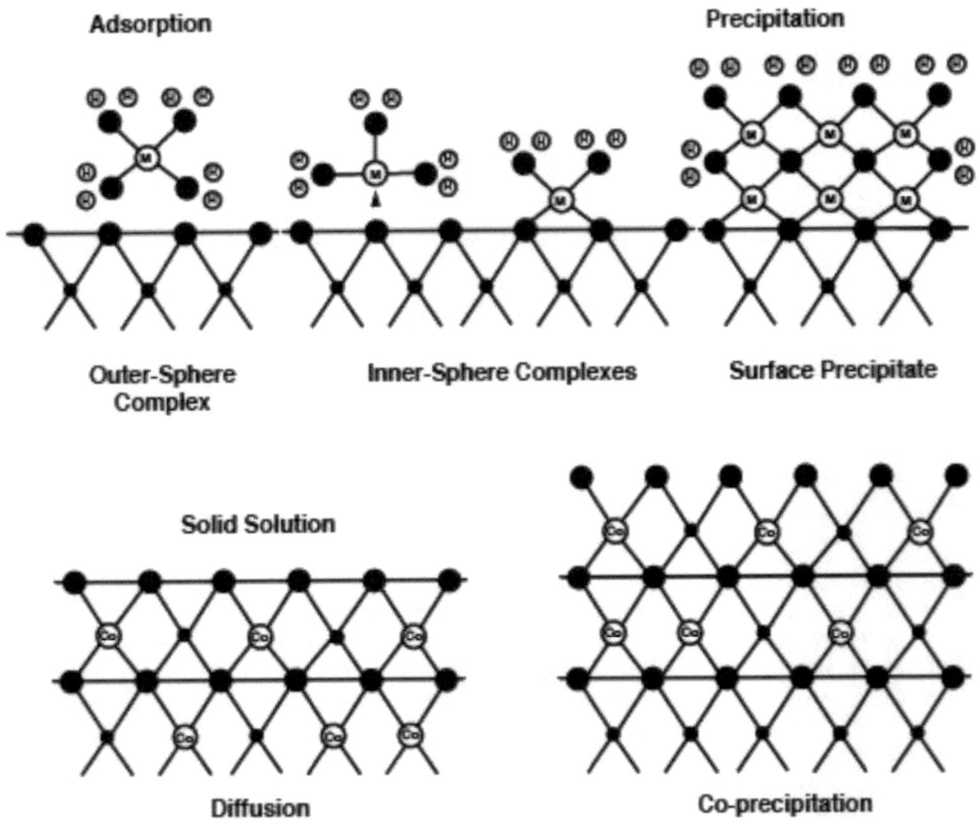

Fig. 5. Possible sorption complexes at the mineral-water interface. M represents aqueous metal ions, and H represents protons. The figures representing diffusion and co-precipitation show two possible modes of Cobalt (Co) absorption in calcite (after Brown et al., 1995).

The underlying material of the oxidation zone in mine tailings has the function of a buffer for the acid and metal bearing solution through sorption and neutralization processes. Once the adsorption and acid-neutralization capacity of the underlying tailings material is exceeded, the mobilized elements may lead to formation of highly metal-bearing acid mine drainage (AMD).

3.8 Microbiological activity

As mentioned before, microbiological activity acts as a catalyst for the oxidation of ferrous to ferric iron, a key process of the acid rock drainage (ARD) problem. Increasing awareness about the role of microorganism in geological processes has lead recently to a new scientific direction called geomicrobiology, e.g. (Banfield & Nealson, 1997; Ehrlich, 1996). The role of microorganisms in the formation of acid mine drainage associated to tailings impoundments and mine waste environments are increasingly a new subject of investigation (Diaby et al., 2007; Johnson & Hallberg, 2003b; Nordstrom & Southam, 1997).

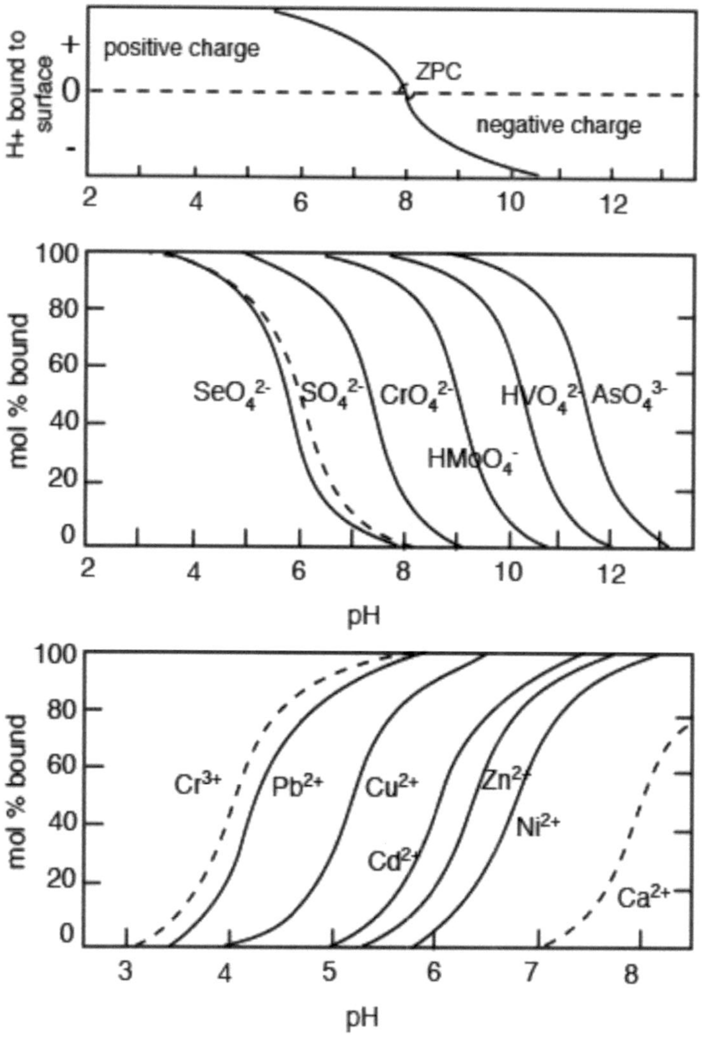

Fig. 6. Adsorption of oxyanions and bivalent cations to Fe(III)hydroxides. With decreasing pH the net surface charge becomes positive due to proton adsorption at the surface. Elements, which are stable at acidic condition as oxyanions become preferentially adsorbed. The adsorption of metals stable as cations increases with pH due to the increasing negative surface charge of the adsorbent. The dashed curves have been calculated (based on data from Dzombak and Morel, 1990; from Stumm and Morgan, 1996).

As mentioned in section 2.4.2.1, *Acidithiobacillus ferrooxidans* has been known to play a key role in sulfide oxidation for 40 years (Singer & Stumm, 1970). These acidophilic chemolithotroph and autotroph bacteria derives cellular carbon from atmospheric CO_2 fixation via the Calvin cycle and obtains energy from the oxidation of Fe(II) or reduced S compounds (H_2S, HS^-, S°, $S_2O_3^{2-}$, SO_3^-). This microbe is also reported to be a facultative H_2-oxidizer and is capable of surviving under anaerobic conditions by utilizing reduced S compounds as an electron donor and Fe(III) as an electron sink (Davis, 1997). *Acidithiobacillus ferrooxidans* is the longest known and most studied organism in acid mine drainage and mine waste environments. Nevertheless, a diverse microbial population of metal-tolerant, neutrophilic to acidophilic sulfide and sulfur-oxidizing Thiobacilli are known so far (Johnson & Hallberg, 2003b; Schippers et al., 1995). *Leptospirilum ferrooxidans* seems to be the dominant genus in some acid environments as reported from Iron Mountain, California (Edwards et al., 1998), mine tailings (Diaby et al., 2007), or leach piles (Rawlings & Johnson, 2007). Also heterotrophic bacteria, green algae, fungi, yeasts, mycoplasma, and amoebae have all been reported from acid mine waters. (Wichlacz & Unz, 1981) isolated 37 acidophilic heterotrophs from acid mine drainage. (Davis, 1997) reports the highest *Acidithiobacillus ferrooxidans* population at the oxidation front, while its heterophobic counterpart *Acidiphilum* spp. show higher population in the upper part of an aged oxidation zone of a mine tailings. (Diaby et al., 2007) have shown that in a porphyry copper tailings impoundment *Leptospirillum ferrooxidans* is the dominant specie at the oxidation front and also with the highest population. Recent data show complex communities structures in pyrite oxidation and bioleaching operation (Halinen et al., 2009; Ziegler et al., 2009). Ehrlich (1996) reported several satellite microorganisms live in close association with *Acidithiobacillus ferrooxidans*. It is nowadays recongnized that an complex ecological interactions control the biogeochemical element cycles in acid environments like the Rio Tinto River, Spain (Gonzalez-Toril et al., 2003). (Barker et al., 1998) reported the increased release of cations from biotite (Si, Fe, Al) and plagioclase (Si, Al) by up to two orders of magnitude by microbial activity compared to abiotic controls. The authors also report the formation of a low pH (3-4) microenvironment associated with microcolonies of bacteria on biotite. These results suggest that in acid rock drainage, tailings and mine waste environments, a complex microbial ecosystem exists, of which the controlling parameters and interactions are poorly understood. This knowledge is not only needed to prevent acid mine drainage and to minimize its hazardous environmental impact, but also to increase metal release in bioleaching operations for more effective metal recovery methods, important aspects for a more sustainable mining approach (Dold, 2008).

3.9 Conclusions

Geochemical conditions in mine waste environments change with time by the exposure of sulfide minerals to atmospheric oxygen and water. Sulfide oxidation is mainly controlled by oxygen and water flux, type of sulfide minerals, type of neutralizing minerals, and the microbial activity. The relation of acid producing processes and neutralizing processes determinates the geochemical Eh-pH conditions and so the mobility of the liberated elements. Thus, it is crucial to determinate the acid producing minerals (primary and secondary) and the acid neutralizing minerals in mine waste in order to predict future geochemical behaviour and the hazardous potential of the material.

Summarizing, it can be stated that for accurate mine waste management assessment, a combination of detailed mineralogical, geochemical, and microbiological studies has to be performed in order to understand and predict the complex geomicrobiological interactions in acid rock drainage formation.

4. References

Acker, J.G. and Bricker, O.P., 1992. The influence of pH on biotite dissolution and alteration kinetics at low temperature. Geochimica et Cosmochimica Acta, 56: 3073-3092.

Ahonen, L. and Tuovinen, O.L., 1994. Solid-Phase Alteration and Iron Transformation in Column Bioleaching of a Complex Sulfide Ore. In: C.N. Alpers and D.W. Blowes (Editors), Environmental Geochimistry of Sulfide Oxidation. ACS Symposium Series, Washington, pp. 79-89.

Alpers, C.N., Blowes, D.W., Nordstrom, D.K. and Jambor, J.L., 1994. Secondary minerals and acid mine-water chemistry. In: J.L. Jambor and D.W. Blowes (Editors), Short course handbook on environmental geochemistry of sulfide mine-waste. Mineralogical Association of Canada, Nepean, pp. 247-270.

Banfield, J.F. and Nealson, K.H. (Editors), 1997. Geomicrobiology. Reviews in Mineralogy, 35. MSA, Washington, DC, 448 pp.

Barker, W.W., Welch, S.A., Chu, S. and Banfield, J.F., 1998. Experimental observations of the effects of bacteria on aluminosilicate weathering. American Mineralogist, 83: 1551-1563.

Baron, D. and Palmer, C.D., 1996. Solubility of jarosite at 4-35°C. Geochimica et Cosmochimica Acta, 60(2): 185-195.

Baumgartner, R., Fontboté, L. and Vennemann, T.W., 2008. Mineral zoning and geochemistry of epithermal polymetallic Zn-Pb-Ag-Cu-Bi mineralization at Cerro de Pasco, Peru. Economic Geology: 493-537.

Bigham, J.M., Schwertmann, U., Traina, S.J., Winland, R.L. and Wolf, M., 1996. Schwertmannite and the chemical modeling of iron in acid sulfate waters. Geochimica et Cosmochimica Acta, 60(12): 2111-2121.

Blowes, D.W. and Ptacek, C.J., 1994. Acid-neutralization mechanisms in inactive mine tailings. In: J.L. Jambor and D.W. Blowes (Editors), Short course handbook on environmental geochemistry of sulfide mine-waste. Mineralogical Association of Canada, Nepean, pp. 271-291.

Blowes, D.W. et al., 1994. Acid-Neutralization Reactions in inactive MIne tailings Impoundments and their Effect on the Transport of dissolved Metals., International Land Reclamation and Mine Drainage Conference and the Third International Conference on the Abatement of Acid Drainage, Pittsburgh, pp. 429-438.

Blowes, D.W., Reardon, E.J., Jambor, J.L. and Cherry, J.A., 1991. The formation and potential importance of cemented layers in inactive sulfide mine tailings. Geochimica et Cosmochimica Acta, 55: 965-978.

Brookins, D.G., 1988. Eh-pH diagrams for geochemistry. Springer, Berlin, 176 pp.

Brown, G.E., Parks, G.A. and O'Day, P.A., 1995. Sorption at mineral-water interfaces: macroscopic and microscopic perspectives. In: D.J. Vaughan and R.A.D. Pattrick (Editors), Mineral Surfaces, The Mineralogical Society Series. Chapman & Hall, London, pp. 129-183.

Bryner, L.C., Walker, R.B. and Palmer, R., 1967. Some factors influencing the biological and non-biological oxidation of sulfide minerals. Transact. Soc. Mining Eng., A.I.M.E, 238: 56-65.

Byrne, P.M. and Beaty, M., 1997. Liquefaction induced displacements. In: P.S. Seco e Pinto (Editor), Seismic behaviour of ground and geotechnical structures. -Proceeding of discussion special technical session on earthquake geotechnical engineering during fourtheenth international conference on soil mechanics and foundation engineering. Balkema, Rotterdam, pp. 185-195.

Carson, C.D., Fanning, D.S. and Dixon, J.B., 1982. Alfisols and ultisols with acid sulfate weathering features in Texas. In: J.A. Kittrick, D.S. Fanning and L.R. Hossner (Editors), Acid sulfide weathering. Soil Science Soc. Am. Pub., Madison, pp. 127-146.

Cornell, R.M. and Schwertmann, U., 2003. The Iron oxides. Wiley-VCH, Weinheim, 664 pp.

Davis, B.S., 1997. Geomicrobiology of the oxic zone of sulfidic mine tailings. In: J.M. McIntosh and L.A. Groat (Editors). Short Course Series. Mineralogical Association of Canada, Nepean, pp. 93-112.

Davis, G.B. and Ritchie, A.I.M., 1986. A model of oxidation in pyritic mine waste: part1 equations and approximate solution. Appl. Math. Modelling, 10: 314-322.

Diaby, N. et al., 2007. Microbial communities in a porphyry copper tailings impoundment and their impact on the geochemical dynamics of the mine waste. Environmental Microbiology, 9(2): 298-307

Dold, B., 2003a. Dissolution kinetics of schwertmannite and ferrihydrite in oxidized mine samples and their detection by differential X-ray diffraction (DXRD). Applied Geochemistry, 18: 1531-1540.

Dold, B., 2003b. Speciation of the most soluble phases in a sequential extraction procedure adapted for geochemical studies of copper sulfide mine waste. Journal of Geochemical Exploration, 80: 55-68.

Dold, B., 2006. Element flows associated with marine shore mine tailings deposits. Environmental Science and Technology, 40: 752-758.

Dold, B., 2008. Sustainability in metal mining: from exploration, over processing to mine waste management. Reviews in Environmental Sciene and Biotechnology, 7: 275-285.

Dold, B. and Fontboté, L., 2001. Element cycling and secondary mineralogy in porphyry copper tailings as a function of climate, primary mineralogy, and mineral processing. Special Issue: Geochemical studies of Mining and the Environment, Journal of Geochemical Exploration, 74(1-3): 3-55.

Dold, B., Wade, C. and Fontbote, L., 2009. Water management for acid mine drainage control at the polymetallic Zn-Pb-(Ag-Bi-Cu) deposit of Cerro de Pasco, Peru. Journal of Geochemical Exploration, 100(2-3): 133-141.

Donnert, D., Eberle, S.H. and Horst, J., 1990. Kinetic studies on the interaction of metals between water and clay mineral. In: J.A.C. Broekaert, S. Gücer and F. Adams (Editors), NATO ASI Series, Metal speciation in the environment, pp. 121-136.

Dutrizac, J.E., Jambor, J.L. and Figures, 2000. Jarosites and their application in hydrometallurgy [Review] Alpers CN, Jambor JL. Sulfate Minerals Crystallography, Geochemistry And Environmental Significance, 9(Prepayment Required).

Dzombak, D.A. and Morel, F.M.M., 1990. Surface complexation modeling - Hydrous ferric oxides. Wiley, New York, 393 pp.

Edwards, K.J., Schrenk, M.O., Hamers, R. and Banfield, J.F., 1998. Microbial oxidation of pyrite: Experiments using microorganisms from extreme acidic environment. American Mineralogist, 83: 1444-1453.

Ehrlich, H.L., 1996. Geomicrobiology. Dekker, New York, 719 pp.

Evangelou, V.P. and Zhang, Y.L., 1995. A review; pyrite oxidation mechanisms and acid mine drainage prevention. Critical Reviews in Environmental Science and Technology, 25(2): 141-199.

Flynn, C.M., 1984. Hydrolysis of inorganic iron(III) salts. Chemical Reviews, 84: 31-41.

Gonzalez-Toril, E., Llobet-Brossa, E., Casamayor, E.O., Ammann, A. and Amils, R., 2003. Microbial ecology of an extreme acid environment, the Rio Tinto River. Appl. Env. Microbiol., 69: 4853-4865.

Halinen, A.K., Rahunen, N., Kaksonen, A.H. and Puhakka, J.A., 2009. Heap bioleaching of a complex sulfide ore. Part I: Effect of pH on metal extraction and microbial composition in pH controlled columns. Hydrometallurgy, 98(1-2): 92-100.

Hodges, C.A., 1995. Mineral resources, environmental issues, and land use. Science, 268: 1305-1312.

Huminicki, D.M.C. and Rimstidt, J.D., 2009. Iron oxyhydroxide coating of pyrite for acid mine drainage control. Applied Geochemistry, 24(9): 1626-1634.

Jambor, J.L., 1994. Mineralogy of sulfide-rich tailings and their oxidation products. In: J.L. Jambor and D.W. Blowes (Editors), Short course handbook on environmental geochemistry of sulfide mine-waste. Mineralogical Association of Canada, Nepean, pp. 59-102.

Jambor, J.L. and Blowes, D.W., 1998. Theory and applications of mineralogy in environmental studies of sulfide-baring mine waste. In: L.J. Cabri and D.J. Vaughan (Editors), Short Course Handbook on Ore and environmental Mineralogy. Mineralogical Society of Canada, Nepean, pp. 367-401.

Jänicke, M. and Weidner, H. (Editors), 1997. National environmental policies. A comparative study of capacity-building. Springer, Berlin, 320 pp.

Jerz, J.K. and Rimstidt, J.D., 2004. Pyrite oxidation in moist air. Geochimica et Cosmochimica Acta, 68(4): 701-714.

Johnson, D.B., 1998. Biodiversity and ecology of acidophilic microorganisms. FEMS Microbiology Ecology, 27: 307-317.

Johnson, D.B., 1999. Importance of microbial ecology in the development of new mineral technologies. In: R. Amils and A. Ballester (Editors), Biohydrometallurgy and the Environment Toward the Mining of the 21st Century. Process Metallurgy. Elsevier, Amsterdam, pp. 645-656.

Johnson, D.B. and Hallberg, K.B., 2003. The microbiology of acidic mine waters. Research in Microbiology, 154(7): 466-473.

Kwong, Y.T.J., 1993. Prediction and prevention of acid rock drainage from a geological and mineralogical perspective, MEND.

Langmuir, D., 1997. Aqueous environmental geochemistry. Prentice Hall, 600 pp.

Malmström, M. and Banwart, S., 1997. Biotite dissolution at 25°C. the pH dependence of dissolution rate and stoichometry. Geochimica et Cosmochimica Acta, 61: 2779-2799.

Majzlan, J., Navrotsky, A. and Schwertmann, U., 2004. Thermodynamics of iron oxides: Part III. Enthalpies of formation and stability of ferrihydrite (~Fe(OH)3), schwertmannite (~FeO(OH)3/4(SO4)1/8), and [epsiv]-Fe2O3 1. Geochimica et Cosmochimica Acta, 68(5): 1049-1059.

Mok, W.M. and Wai, C.M., 1994. Mobilization of arsenic in contaminated river waters. In: J.O. Nriagu (Editor), Arsenic in the environment. Part I Cycling and characterization. John Wiley Interscience, New York, pp. 99-108.

Morin, A.K. and Cherry, J.A., 1986. Trace amoumts of siderite near a uranium-tailings impoundments, Elliot Lake, Ontario, and its implications in controlling contaminant migration in a sand aquifer. Chemical Geology, 56: 117-134.

Moses, C.O., Kirk Nordstrom, D., Herman, J.S. and Mills, A.L., 1987. Aqueous pyrite oxidation by dissolved oxygen and by ferric iron. Geochimica et Cosmochimica Acta, 51(6): 1561-1571.

Nesbitt, H.W. and Jambor, J.L., 1998. Role of mafic minerals in neutralizing ARD, demonstrated using a chemical weathering methodology. In: L.J. Cabri and D.J. Vaughan (Editors), Modern approaches to ore and environmental mineralogy, Short Course Handbook, pp. 403-421.

Nicholson, R.V. and Scharer, J.M., 1994. Laboratory studies of pyrrhotite oxidation kinetics. In: C.N. Alpers and D.W. Blowes (Editors), Environmental Geochimistry of Sulfide Oxidation. ACS Symposium Series. American Chemical Society, Washington, pp. 14-30.

Nordstrom, D.K., 1982. Aqueous pyrite oxidation and the consequent formation of secondary iron minerals. In: J.A. Kittrick and D.S. Fanning (Editors), Acid sulfate weathering. Soil Sci. Soc. Am. J., pp. 37-56.

Nordstrom, D.K. and Alpers, C.N., 1999. Geochemistry of acid mine waste. In: G.S. Plumlee and M.J. Logsdon (Editors), The environmental geochemistry of ore deposits. Part A: Processes, techniques, and health issues. Reviews in Economic Geology, pp. 133-160.

Nordstrom, D.K., Alpers, C.N., Ptacek, C.J. and Blowes, D.W., 2000. Negative pH and extremely acidic mine waters from Iron Mountain, California. Environ. Sci. Technol., 34: 254-258.

Nordstrom, D.K., Jenne, E.A. and Ball, J.W., 1979. Redox equilibria of iron in acid mine waters, Chemical modeling in aqueous systems. ACS Symposium Series. American Chemical Society, pp. 51-79.

Nordstrom, D.K. and Southam, G., 1997. Geomicrobiology of sulfide mineral oxidation. In: J.F. Banfield and K.H. Nealson (Editors), Geomicrobiology. Reviews in Mineralogy. Mineralogical Society of America, pp. 361-390.

Norris, P.R. and Johnson, D.B., 1998. Acidophilic microorganisms. In: K. Horikoshi and W.D. Grant (Editors), Extremophiles: Microbial Life in Extreme Environments. John Wiley, New York, pp. 133-154.

Parks, G.A., 1990. Surface energy and adsorption at mineral-water interfaces: an introduction. In: M.F. Hochella and A.F. White (Editors), Mineral-Water Interface Geochemistry. Reviews in Mineralogy, pp. 133-175.

Plumlee, G.S., 1999. The environmental geology of mineral deposits. In: G.S. Plumlee and M.J. Logsdon (Editors), The environmental geochemistry of ore deposits. Part A: Processes, techniques, and health issues. Reviews in Economic Geology, pp. 71-116.

Rawlings, D.E. and Johnson, D.B., 2007. Biomining. Springer, Berlin Heidelberg, 314 pp.

Rimstidt, J.D., Chermak, J.A. and Gagen, P.M., 1994. Rates of Reaction of Galena, Spalerite, Chalcopyrite, and Asenopyrite with Fe(III) in Acidic Solutions. In: C.N. Alpers and D.W. Blowes (Editors), Environmental Geochimistry of Sulfide Oxidation. ACS Symposium Series. American Chemical Society, Washington, pp. 2-13.

Rimstidt, J.D. and Vaughan, D.J., 2003. Pyrite oxidation: a state-of-the-art assessment of the reaction mechanism. Geochimica et Cosmochimica Acta, 67(5): 873-880.

Ritcey, G.M., 1989. Tailings management. problems and solutions in the mining industry. r. Elsevier, Amsterdam, 970 pp.

Ritchie, A.I.M., 1994. Sulfide oxidation mechanisms: controls and rates of oxygen transport. In: J.L. Jambor and D.W. Blowes (Editors), Short course handbook on environmental geochemistry of sulfide mine-waste. Mineralogical Association of Canada, Nepean, pp. 201-244.

Sato, M., 1960. Oxidation of sulfide ore bodies, II. Oxidation mechanisms of sulfide minerals at 25°C. Economic Geology, 55: 1202-1231.

Schippers, A., Hallmann, R., Wentzien, S. and Sand, W., 1995. Microbial diversity in uranium mine waste heaps. Appl. Environ. Microbiol., 61: 2930-2935.

Schneider, W. and Schwyn, B., 1987. The hydrolysis of iron in synthetic, biological, and aquatic media. In: W. Stumm (Editor), Aquatic surface chemistry. Wiley-Interscience Publication, New York, pp. 519.

Schwertmann, U., 1964. Differenzierung der Eisenoxide des Bodens durch Extraktion mit Ammoniumoxalat Lösung. Zeitschrift für Pflanzenernährung und Bodenkunde, 105: 194-202.

Singer, P.C. and Stumm, W., 1970. Acid mine drainage: the rate-determining step. Science, 167: 1121-1123.

Strömberg, B. and Banwart, S., 1994. Kinetic modelling of geochemical processes at Aitik mining waste rock site in northern Sweden. Applied Geochemistry, 9: 583-595.

Stumm, W. and Morgan, J.J., 1996. Aquatic chemistry. Wiley, New York, 1022 pp.

Stumm, W. and Sulzberger, B., 1992. The cycling of iron in natural environments: Considerations based on laboratory studies of heterogeneous redox processes. Geochimica et Cosmochimica Acta, 56: 3233-3257.

Suter, D., Siffert, C., Sulberger, B. and Stumm, W., 1988. Catalytic dissolution of iron(III)(hydr)oxides by oxalic acid in the presence of Fe(II). Naturwiss., 75: 571-573.

Sylva, R.N., 1972. The hydrolysis of iron(III). Rev. Pure and Appl. Chem., 22: 115-132.

Weiss, N.L. (Editor), 1985. SME mineral processing handbook. SME, 2144 pp.

White III, W.W., Lapakko, K.A. and Cox, R.L., 1999. Static-test methods most commonly used to predict acid mine drainage: Practical Guidlines for use and interpretation. In: G.S. Plumlee and M.J. Logsdon (Editors), The environmental geochemistry of ore deposits. Part A: Processes, techniques, and health issues. Reviews in Economic Geology, pp. 325-338.

Wichlacz, P.L. and Unz, R.F., 1981. Acidophilic, heterotrophic bacteria of acid mine drainage waters. Appl. Environ. Microbiol., 41: 1254-1261.

Ziegler, S., Ackermann, S., Majzlan, J. and Gescher, J., 2009. Matrix composition and community structure analysis of a novel bacterial pyrite leaching community. Environmental Microbiology, 11(9): 2329-2338.

11

Synthetic Aggregates Produced by Different Wastes as a Soil Ameliorant, a Potting Media Component and a Waste Management Option.

Guttila Yugantha Jayasinghe and Yoshihiro Tokashiki
Department of Environmental Science and Technology, Faculty of Agriculture, University of the Ryukyus, Senbaru-1, Nishihara-Cho, Okinawa (903-0213), Japan

1. Introduction

In most developed and developing countries with increasing population, prosperity and urbanization, one of the major challenges for them is to collect, recycle, treat and dispose of increasing quantities of solid waste and wastewater. It is now well known that waste generation and management practices have increased several alarming issues on the socio-economics, human health, aesthetics and amenity of many communities, states, and nations around the world (Meyers et al., 2006; Louis, 2004). Industrialized economies extract vast quantities of natural resources from the environment to provide modern amenities and commodities. On the other hand, pollutants associated with the production and consumption of commodities, as well as post-consuming commodities, go back into the environment as residues (Moriguchi, 1999). Although varying in degree and intensity, the solid waste problem around the world is exacerbated by limited space and dense populations (Melosi, 1981). The problem of collecting, handling and disposing of wastes is dealt with using different techniques and approaches in different regions. A waste management hierarchy based on the most environmentally sound criteria favors waste prevention/minimization, waste re-use, recycling, and composting. In many countries, a large percentage of waste cannot presently be re-used,re-cycled or composted and the main disposal methods are land filling and incineration. In addition, traditionally, managing domestic, industrial and commercial waste consisted of collection followed by disposal, usually away from urban activity, which could be waterways, Open ocean or surface areas demarcated for the purpose viz. landfills. With the increased volume and variety of hazards posed by new waste products, the situation has exceeded its saturation point at many localities (McCarthy, 2007). In 2006 the USA land filled 54% of solid wastes, incinerated 14%, and recovered, recycled or composted the remaining 32% (EPA, 2008). The percentage of solid waste disposed at landfills accounted for 3% in Japan (2003), 18% in Germany (2004), 36% in France (2005), 54% in Italy (2005) and the USA (2005), and 64% in the UK (2005). As legislation becomes more stringent and land filling becomes less cheap option. For example, there has been a significant reduction in the amount of wasteland filled in the UK and Italy. In 1995, Italy land filled 93% of solid waste, and the UK 83%. Recent studies have revealed that waste disposal processes have considerable impacts on climate change due to the associated greenhouse gases (GHGs) emission (Elena, 2004; Sandulescu, 2004; USEPA, 2002). Land filling processes are found to be the largest anthropogenic source of CH_4 emission in the United States. In 2004, there were 140.9 Tg of CO_2 equivalent of CH_4 (approximately 25% of the United States' annual CH_4 emission) emitted from the landfills, which shared 2.65% of the national global-warming damage. In addition, 19.4 and 0.5 Tg of CO_2 equivalent of CO_2 and N_2O were, respectively, released from the combustion processes (USEPA, 2006). These evidences show that waste disposal systems are one of the most significant contributors to potential climate change, as the associated-emission cannot be effectively mitigated under current management conditions. Moreover, Incineration is also cannot be recommended as an efficient method since it is also creating toxic gases and GHGs. In addition, wide range of waste materials (sewage sludge, industrial waste) is increasingly spread on agricultural land as soil amendments. These undoubtedly produce a number of positive effects on soil quality, but also raise concern about potential short-term (e.g. pathogen survival) and long-term effects (e.g. accumulation of heavy metals). Climate change will also become a major incentive to the use of biosolids on agricultural land, especially in regions where longer periods of low rainfall and mean higher temperatures are expected. In many parts of the world (e.g. Europe, USA) agricultural soils receive large volumes of soil amendments. Approximately 5.5 million dry tones of sewage sludge are used or disposed of annually in the United States and approximately 60% of it is used for land application (NRC, 2000). The application of biosolids to soil is likely to increase as a result of the diversion of waste away from landfill sites, and due to increasing cost of artificial fertilizers (UNEP, 2002; Epstein, 2003). Simply application of waste as an amendment to agricultural lands made some environmental problems such as air pollution due to tiny particles of coal fly ash (CFA). Therefore, it is worthwhile to find out alternative methods for waste disposal. Consequently, unconventional synthetic aggregates were produced from different waste materials (sewage sludge, paper waste, oil palm waste, sugarcane trash, starch waste, CFA, wood chips, coir dust, cattle manure compost, chicken manure compost etc…) to utilize them in agriculture as a soil amendment, fertilizer support, and potting media for containerized plant cultivation (Jayasinghe & Tokashiki, 2006; Jayasinghe et al., 2005, 2008, 2009 a,b,c,d,e,f,g). These synthetic aggregates proved that they can be utilized in agriculture very effectively. Moreover, these kinds of unconventional synthetic aggregate production have not much been reported in the literature. Therefore, this chapter describes the production, characterization and different utilization methods of synthetic aggregates in agriculture.

2. What is a Synthetic Aggregate (SA)?

Aggregate structure is schematically shown in Figure 1. It is composed with rigid or composite materials, fibrous materials and a binder.

2.1 Rigid or composite materials

Sewage sludge, sugarcane trash, wood chip, CFA, compost, soil etc. can be regarded as rigid materials. The rigid materials give the rigidity and the strength of the aggregate by enmeshing into fibrous matrix. Figure 2 shows the scanning electron microscopic (SEM) image of a coal fly ash paper waste aggregate, which is showing the rigid CFA particles are enmeshed into the fibrous paper waste matrix by the binder.

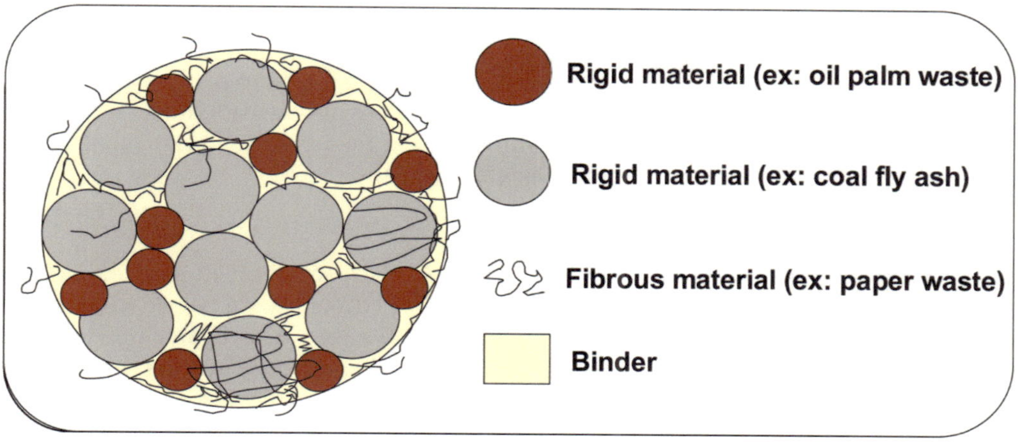

Fig. 1. Schematic diagram of the synthetic aggregate.

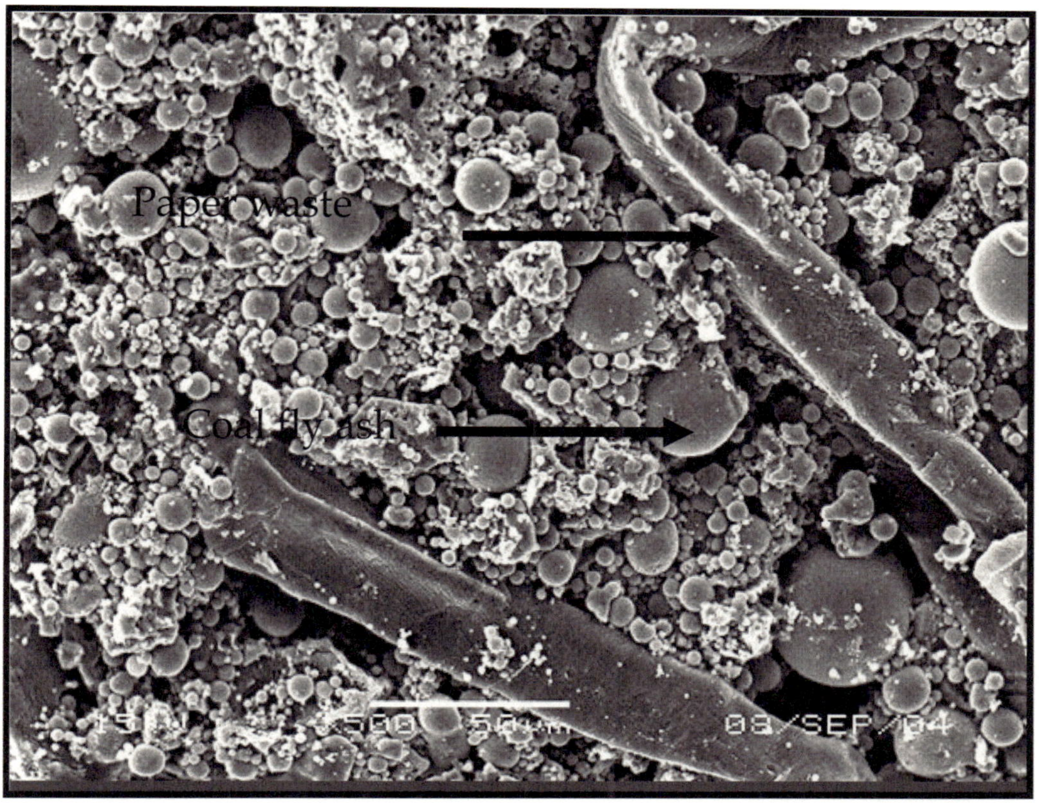

Fig. 2. Scanning electron micrograph of a coal fly ash-paper waste synthetic aggregate.

2.2 Fibrous materials

The formation of aggregate requires a matrix to adhere the rigid particles. Then this matrix can form the aggregate structure by binding the rigid particles into the matrix by the binder. Paper waste, coco fiber, wheat and rice straw and oil palm fiber can be used as the fibrous materials. Figure 2 shows the porous paper waste matrix, which provides the binding sites to the CFA particles. Porous spaces can be observed within the aggregate, which can improve the aeration and the water holding capacity of the aggregates as a growth substrate (Jayasinghe et al., 2007). Fibrous materials in the aggregates also assist to increase the micro pores in the aggregate-soil amendment mixtures during the humification of aggregates by microbes after mixing them into the soil as an amendment.

2.3 Binder

The formation of aggregate requires both physical rearrangement of particles and the stabilization of the new arrangement. Therefore effective binder should be added in order to obtain stable aggregates. Several binding mechanisms exist between organic polymers and mineral surfaces to provide stable aggregates. Organic polymers have been used quite effectively to stabilize soil structure in recent years. Many researchers have shown that the application of polyacrylamide maintained high infiltration rate during rainfall and reduced soil surface sealing and runoff soil losses (Ben-Hur & Keren, 1997; Sojka et al., 1998; Green et al., 2000). Polysaccharides added to soil as soil conditioners improve soil's physical properties that are important for plant growth and increase soil's resistance against disruptive forces and erosion. Organic polymers have been used quite effectively to stabilize soil structure in recent years. Polysaccharides stabilize soil aggregates because of their contribution as cements and glues. (Taskin etal., 2002).There is a considerable amount of starch which is a polysaccharide coming out as waste material from Okinawa flour industry (Okinawa, Seifun ltd). Utilization of the starch waste is currently under the potential capacity. Therefore, the starch waste was utilized as an organic binder for the synthetic aggregate production. In addition, several inorganic binders can be used to produce synthetic aggregates. Acryl resin emulsion binder EMN-coat /21 and Calcium hydroxide with calcium sulfate can also be used as the binder to produce aggregates.

2.4 Production of synthetic aggregates

Production process of aggregates is given in Figure3. EIRICH mixer, Ploughshare mixer or Pelleger machine can be used for the production of heterogeneous aggregates. Heterogeneous aggregates means the aggregates containing different particle sizes. Pelleter machine can be utilized to produce homogenous (same size) aggregates. EIRICH mixer was used for small scale aggregate production and pelleger machine and pelleter was used in major scale aggregates production. Different proportions of raw materials were mixed in the pelleger or EIRICH mixer for 1-3 minutes. Then binder was added and mixed for another 1-2 minutes. Finally whole mixture was mixed for another 2-5 minutes in high speed rotation to form aggregates. Raw materials with binder mixture were inserted to the pelleter machine for the production of homogenous aggregates. Moreover, diameter and the length of the aggregates can be adjusted according to the requirement.

2.5 Different types of aggregates with various types of wastes.

Different types of aggregates can be developed with the available waste in the site or area. Some of the developed aggregates from different wastes are described below (Figure 4). Basically aggregates can be divided into two types.

1. Heterogeneous aggregates

   These aggregates contain different sized aggregates. Following are some of the heterogeneous aggregates developed from various materials.

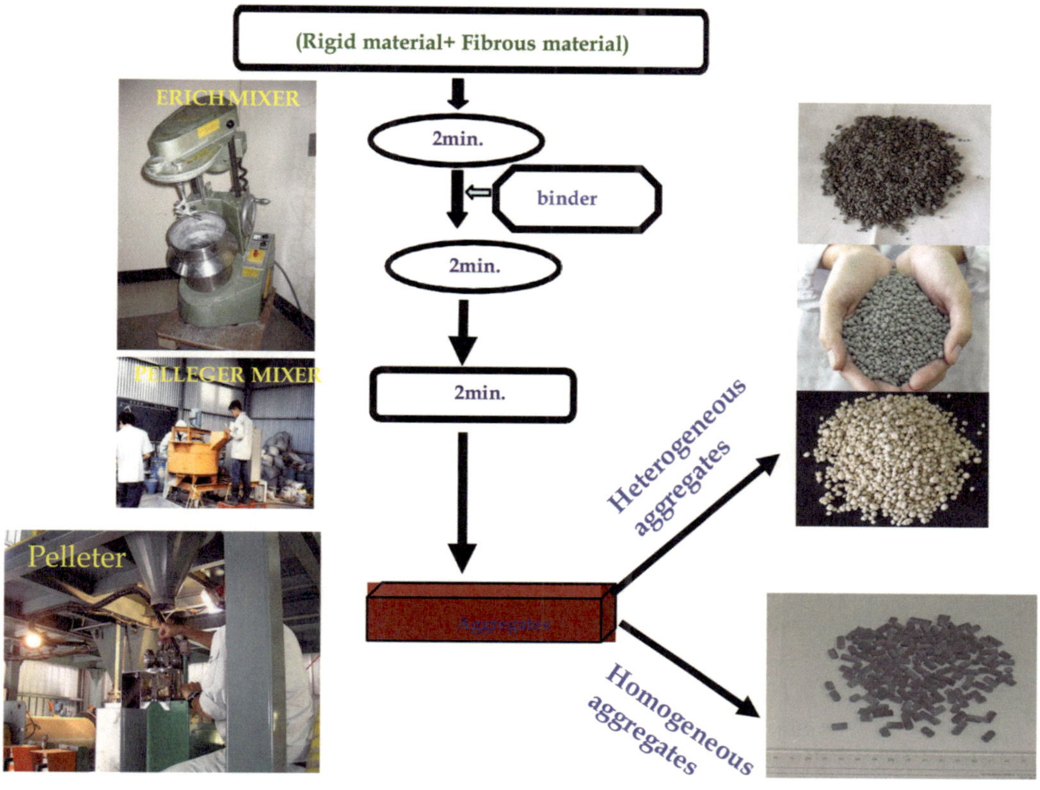

Fig. 3. Production process of different types of aggregates.

i. Coal fly ash based aggregates
These aggregates were developed from CFA, paper waste or oil palm waste with organic or inorganic binders (Figure 4a).

ii. Soil aggregates
These were developed from low productive acidic red soil with paper waste, coco fiber, or oil palm waste (Figure 4b) with organic or inorganic binders.

iii. Acid soil-coal fly ash aggregates
These were developed by acid soil and the coal fly ash with paper waste, sewage sludge (SS), CFA with organic or inorganic binder. (Figure 4 c)

iv. Sewage sludge based aggregates
These aggregates were developed from sewage sludge and zeolite with an inorganic binder. (Figure 4d)

v. Compost based aggregates
These were produced from different types of composts and soil with organic or inorganic binders (Figure 4e)

2. Homogenous aggregates
These aggregates have same sized aggregates and pelleter machine was used for the production of these aggregates.
These aggregates are called as synthetic pellet aggregates. Coal fly ash (CFA), soil, compost, paper waste, coco fiber, oil palm waste, sewage sludge and organic or inorganic binders can be utilized as raw materials for these types of aggregates. (Figure 4f)

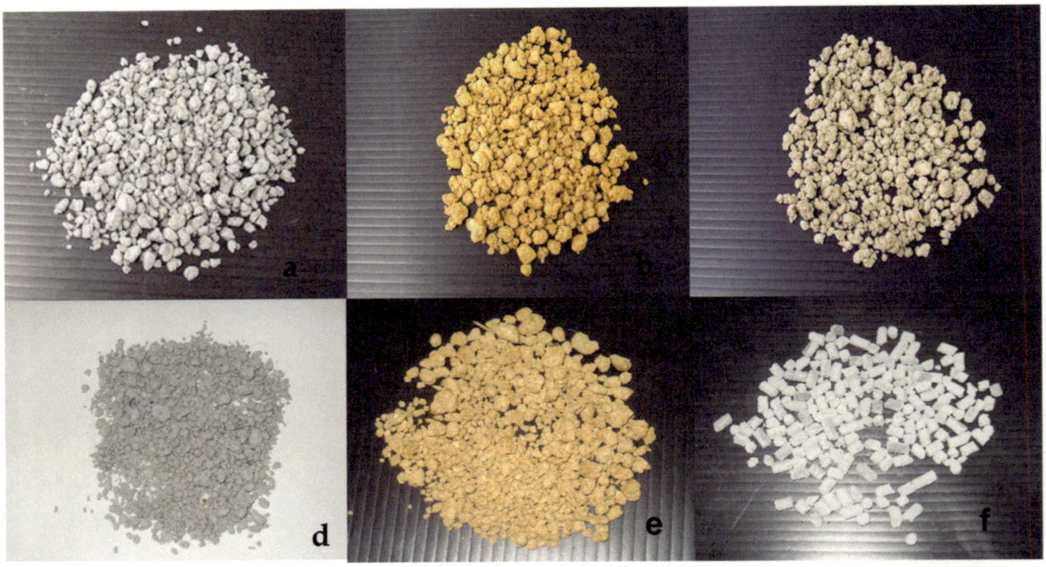

Fig. 4. Different types of aggregates produced from different materials.
(a) coal fly ash paper waste aggregates, (b) soil aggregates,
(c) acid soil coal fly ash aggregates, (d) sewage sludge based aggregates,
(e) compost based aggregates, (f) pellet aggregates (homogeneous aggregates)

3. Physical and chemical properties of synthetic aggregates

3.1 Physical properties

Particle size distributions of synthetic aggregates developed by different materials are given in the Table 1. Particle size distribution of a substrate is important because it determines pore space, air and water holding capacities (Raviv et al., 1986). Mean distribution of the aggregate media showed that fraction between 5.60 and 2.00 mm was the most abundant fraction in all types of synthetic aggregates (Table 1). An excess of fines in a substrate clogs pores, increases non-plant-available water holding capacity and decreases air filled porosity (Spiers & Fietje, 2000). Therefore, these synthetic aggregates which are having higher percentage of larger- sized particles can be utilized to enhance the properties of problematic soils having higher finer particles to improve its porosity and hydraulic conductivity. Synthetic aggregates developed from low productive acid soil and paper waste addition to the problematic grey soil in Okinawa, Japan significantly enhanced the particles >2.00 mm and hence the hydraulic conductivity and porosity were significantly increased (Jayasinghe et al., 2009d). All of synthetic aggregates shown in the Table 1 are heterogeneous which are having different sized aggregates. Synthetic pellet aggregates can be developed with single particle sized diameter which are called as homogenous aggregates. These pellet aggregates can be produced with the required diameter as a tailor made production. In addition, aggregate diameter of heterogeneous aggregates depends upon the material type and quantity, binder type and quantity and the mixing time. Therefore, suitable particle sizes of heterogeneous synthetic aggregates can be designed according to the requirement. Synthetic aggregates particle sizes are varying with the required situation. For an example particle sizes for a potting medium are different from particle sizes required as a soil ameliorant. As a potting medium relatively higher percentage of smaller sized-diameter particles should be utilized to improve the potting media characteristics.

Aggregate type	>5.60 mm	5.60-3.35 mm	3.35-2.00 mm	2.00-1.00 mm	1.00-0.50 mm	>0.50 mm
A	25.61	30.19	28.32	8.54	5.20	2.14
B	10.77	19.85	36.98	13.15	7.91	11.34
C	23.86	28.14	25.74	9.44	8.09	4.73
D	6.34	16.92	16.98	19.44	21.23	19.09
E	26.20	30.28	27.54	8.73	5.02	2.23
F	7.10	15.21	18.77	15.91	22.87	20.14

A: coal fly ash paper waste aggregates with starch binder (Jayasinghe et al., 2009b),
B: acid soil aggregates with starch binder (Jayasinghe et al., 2008),
C: acid soil coal fly ash aggregates with starch binder, (Jayasinghe et al., 2008),
D: acid soil compost aggregates with inorganic binder,
E: coal fly ash aggregates with inorganic binder (Jayasinghe et al., 2009a),
F: sewage sludge aggregates with inorganic binder.

Table 1. Particle size distribution of synthetic aggregates.

Bulk density, particle density, hydraulic conductivity, water holding capacity and aggregate strength of the synthetic aggregates are given in the Table2. Bulk density of a substrate gives a good indication of porosity, which determines the rate at which air and oxygen can move through the substrate. Bulk density values of all substrate given in the table showed low values compared to red soil (1.26 gcm^{-3}) in Okinawa Japan. These low values are due to the coal fly ash, paper waste, and sewage sludge in the developed synthetic aggregates. It is also evident that there were significant differences between bulk density values of aggregates produced with different coal fly ash additions to red soil (Jayasinghe et al., 2005). The particle density of the synthetic aggregates was also low compared to the red soil. Red soil gave a particle density of 2.61 gcm^{-3}. Hydraulic conductivity of a substrate is a measure of the ability of air and water to move through it. Hydraulic conductivity is influenced by the size, shape and continuity of the pore spaces, which in turn depend on the bulk density, structure and the texture. Hydraulic conductivity of the aggregates showed higher values compared to the red soil and grey soil studied in Okinawa, Japan. The red soil and grey soils showed hydraulic conductivity values of 6.62 $x10^{-5}$ and 6.67 $x10^{-5}$, respectively. The water holding capacity of the synthetic aggregates given in the table varied between 0.59 and 0.68 $kgkg^{-1}$, which are increased values compared to the red soil (0.48 $kgkg^{-1}$) in Okinawa Japan (Jayasinghe et al., 2009e). Aggregate strength of the produced aggregates varied between 2.58-4.01 $kgcm^{-2}$. Synthetic aggregates developed by using coal fly ash, paper waste, and starch waste gave an average aggregate strength in the range of 2.05-3.58 $kgcm^{-2}$, which can be considered as higher aggregate strengths (Jayasinghe et al., 2005, 2006, 2008). Higher aggregate strengths indicate resistance of the aggregate to the erosion. Therefore, these synthetic aggregates can withstand to erosion compared to soil particles.

3.2 Chemical properties

Chemical properties of the different types of synthetic aggregates are given in the Table 3. It is evident that pH of aggregates were varied in a wide range from 4.57 to 10.72. A, C, E and

G aggregates having higher pH values were produced by using CFA as a material in the aggregates. The original pH of the CFA used to produce synthetic aggregates was varied in the range of 11.36-11.80. Therefore, CFA aggregates gave alkaline pH values. The hydroxide and carbonate salts in CFA gave one of its principle beneficial chemical characteristics, the ability to neutralize acidity in soils (Pathan et al., 2003). Therefore, these alkaline synthetic aggregates can be used as a buffer material to neutralize the acidic problematic soils. Jayasinghe et al., (2006) reported that 25% of synthetic CFA based aggregates addition increased the acidic pH (4.62) of red soil into 6.25.Type B aggregates were developed from acidic red soil with paper wastes showed acidic pH of 4.57 due to acidic soil. Type D aggregates were produced from acidic red soil with cattle manure compost which neutralized the acidic pH of the red soil and gave a pH of 6.40. Type F aggregates gave a pH of 7.58 due to the alkaline sewage sludge (pH=7.72) in the aggregates. It is evident that the pH of the aggregates depends on the materials which were used to form the aggregates. Aggregates showed high electrical conductivity (EC) except type B due to high essential and non essential elements in the aggregates. Type B aggregates produced from red soil and paper waste, which did not contain much element concentrations, gave the lowest EC. But coal fly ash had high concentrations of different elements, which subsequently raised the EC of the CFA based aggregates. The EC and metal content of soil increases with increasing amount of CFA application (Sikka & Kansal, 1994). Aggregates developed from sewage sludge (SS) also showed high EC due to the presence of high concentrations of elements in the sewage sludge (SS). Gil et al., (2008) reported that SS was characterized by higher EC. The High EC in SS may be due to presence of high concentrations of different types of elements.

Aggregate type	Bulk density (gcm^{-3})	Particle density (gcm^{-3})	Hydraulic conductivity (cms^{-1})	Water holding capacity ($kgkg^{-1}$)	Aggregate strength ($kgcm^{-2}$)
A	0.56	2.48	$2.80x10^{-2}$	0.62	3.91
B	0.87	2.44	$1.87x10^{-2}$	0.63	2.58
C	0.80	2.20	$3.74x10^{-2}$	0.68	3.06
D	0.64	2.48	$2.80x10^{-2}$	0.67	3.76
E	0.64	2.31	$1.87x10^{-2}$	0.59	3.88
F	0.54	2.08	$2.24x10^{-2}$	0.61	4.01
G	0.52	2.20	$2.80x10^{-2}$	0.60	3.92

A: coal fly ash paper waste aggregates with starch binder,
B: acid soil aggregates with starch binder,
C: acid soil coal fly ash aggregates with starch binder,
D: acid soil compost aggregates with inorganic binder,
E: coal fly ash aggregates with inorganic binder,
F: sewage sludge aggregates with inorganic binder,
G: synthetic pellet aggregates (diameter is 10 mm).

Table 2. bulk density, particle density, hydraulic conductivity, water holding capacity and aggregates strength of the synthetic aggregates.

	A	B	C	D	E	F	G
pH	9.82	4.57	9.71	6.40	10.72	7.58	9.28
EC(mSm^{-1})	96.16	6.36	57.50	48.76	90.40	156.26	80.76
C ($g\ kg^{-1}$)	120.82	85.40	66.21	101.12	55.22	291.80	113.61
N ($g\ kg^{-1}$)	0.71	0.40	0.40	1.06	0.42	29.10	0.62
P ($g\ kg^{-1}$)	0.11	0.08	0.05	0.46	0.06	14.65	0.20
Na ($g\ kg^{-1}$)	0.87	0.24	0.44	0.71	0.78	0.54	0.88
K ($g\ kg^{-1}$)	1.51	0.18	0.76	2.34	1.56	0.62	1.61
Mg ($g\ kg^{-1}$)	0.72	0.38	0.47	0.87	0.73	4.12	0.91
Ca ($g\ kg^{-1}$)	3.34	1.10	2.31	2.12	37.25	65.64	3.18
B ($mg\ kg^{-1}$)	16.86	0.12	10.33	0.42	19.34	0.51	12.17
Mn ($mg\ kg^{-1}$)	15.82	20.28	18.73	24.14	19.20	109.66	14.88
Cu ($mg\ kg^{-1}$)	18.47	13.21	16.22	22.66	18.50	188.02	19.12
Zn ($mg\ kg^{-1}$)	34.63	21.35	28.93	32.12	34.60	485.07	31.98
Cr ($mg\ kg^{-1}$)	7.62	1.21	5.45	0.98	7.60	34.42	7.02
Cd ($mg\ kg^{-1}$)	ND	ND	ND	ND	ND	0.40	ND
Se ($mg\ kg^{-1}$)	ND	ND	ND	ND	ND	ND	ND
Pb ($mg\ kg^{-1}$)	7.56	3.01	5.88	3.66	7.60	26.33	8.02
As ($mg\ kg^{-1}$)	ND	ND	ND	ND	ND	ND	ND

A: coal fly ash paper waste aggregates with starch binder,
B: acid soil aggregates with starch binder,
C: acid soil coal fly ash aggregates with starch binder,
D: acid soil compost aggregates with inorganic binder,
E: coal fly ash aggregates with inorganic binder,
F: sewage sludge aggregates with inorganic binder,
G: synthetic pellet aggregates (diameter is 10 mm).
EC: electrical conductivity, ND: not detected.

Table 3. Chemical properties of synthetic aggregates.

Carbon (C) content of aggregates also varied in a greater range and depends on the material type in the aggregate. The N content of the aggregate types of A, B, C, E and G showed low N amount. But D and F gave high N content. Aggregate D contains cattle manure compost while F contains sewage sludge. Moreover, aggregates enriched with N, P and K can be developed by adding respective N, P and K chemical fertilizer as a material to produce aggregates. All of the aggregates gave low phosphorous (P) content except the type E since it is composed with high P containing SS. Aggregates having CFA, compost and SS (A, C, D, E and F) gave high concentrations of Na, K, Mg and Ca in the aggregates. Chemically, 90–99% of CFA is comprised of silicon (Si), aluminum (Al), Ca, Mg, Na and K (Adriano et al., 1980). Aggregates developed from coal fly ash gave high boron (B) content compared to other aggregates. This is due to high B content of the CFA. CFA contains significant levels of B (Lee et al., 2008).

Heavy metal concentrations of the different aggregates are given in Table 3. Selenium (Se) and Arsenic (As) were not detected in any aggregates, and Cadmium (Cd) was detected only in F. The copper (Cu), chromium (Cr), manganese (Mn), zinc (Zn) and lead (Pb) concentrations were generally well below the maximum pollutant concentrations of individual metals for land application suggested by the US Environmental Protection Agency (USEPA, 1993). The maximum pollutant concentrations of individual heavy metal content for land application of sewage sludge given by the US Environmental Protection Agency are (all in mg/kg); As 41, Cr 1200, Cu 1500, Zn 2800, Pb 300, Cd 39 and Se 36, respectively (USEPA, 1993). Furthermore, average concentrations of heavy metals reported in uncontaminated soils are (all in $mgkg^{-1}$); 6 As, 70 Cr, 30 Cu, 90 Zn, 35 Pb and 0.35 Cd, respectively (Adriano, 2001). Though the concentrations of heavy metals were below the uncontaminated soil values and not alarming, there should be routine inspections to ensure that heavy metal concentrations remain within safe limits.

4. Aggregate utilization

4.1 Synthetic aggregates as a soil ameliorant to problematic soils

4.1.1 Synthetic aggregates as a soil ameliorant to low productive acidic red soil.

4.1.1.1 Coal fly ash paper waste starch binder aggregates

Widely spread red soil ("Kunigami Mahji") in sub-tropical Okinawa, Japan, is not suitable for crop production due to its poor physical (Tokashiki et al., 1994) and chemical properties, such as its acidic nature, low organic matter content, and poor nutrient availability (Kobayashi & Shinagawa, 1966; Hamazaki, 1979).Therefore, CFA paper waste aggregates developed with CFA, paper waste and starch binder were used as a soil ameliorant to improve the low productive acidic soil. Aggregates were produced by combining CFA and paper waste using an Eirich mixer (R-02M/C27121) with starch binder.500 g of coal fly ash and 50 g of paper waste were mixed in the Eirich mixer by adding 250 ml of starch binder to produce aggregates. Developed aggregates were used as a soil ameliorant to low productive acidic soil with the objective of enhancing the soil physical and chemical properties to improve the growth and development of Komatsuna (*Brasica rapa*) which is a popular vegetable in Japan. The different amendment rates of the experiment are given in Table 4.

Treatments	Description
T1	Aggregates only
T2	75 % of aggregates
T3	50% of aggregates
T4	25% of aggregates
T5	10% of aggregates
T6	Acidic red soil only.

Table 4. Different treatments were used under the study.

4.1.1.1.1 Influence of aggregate addition to red soil on the growth and development of Komatsuna

Aggregate addition of 25% with acidic red soil ("Kunigami Mahji") as a soil amendment, favorably improved Komatsuna yield by giving the highest significant increase in plant height, fresh and oven dry yield over other treatments (Table 5). Treatments of 10% of aggregate addition gave the second highest average values of yield. Aggregates only (T1) and acid soils only (T6) showed the lowest mean values of height and yield (Table 5). Therefore, aggregate addition to the soil as a soil amendment improves the crop production in comparison with acidic red soil. Aggregate addition percentages of 50% and 75% did not show any significant difference between them, but, significantly differed from the treatments of 25% of aggregate addition and acidic red soil only. Aggregate addition modified the acidic pH conditions to nearly neutral conditions at the 10% and 25% percentages. Therefore, synthetic aggregates, which were produced from coal fly ash, buffer the acidic red soils, while forming conducive crop growth environment by binding soil particles with coal fly ash aggregates, which improves the crop growth and soil physico - chemical properties.

Treatments	Dry weight ($gpot^{-1}$)	Fresh weight ($gpot^{-1}$)	Height (cm)
T1	0.25^{d}	1.50^{e}	5.02^{d}
T2	2.33^{c}	12.50^{d}	14.13^{c}
T3	2.92^{c}	17.00^{c}	16.16^{c}
T4	8.10^{a}	48.00^{a}	24.21^{a}
T5	7.13^{b}	45.00^{b}	21.08^{b}
T6	0.30^{d}	2.00^{e}	6.11^{d}

(Means followed by the different superscript letter in the same column differed significantly according to Duncan's multiple range test (P=0.05).

Table 5. Influence of aggregates as a soil ameliorant on crop growth and yield of Komatsuna.

4.1.1.1.2 Effect of aggregate addition on soil characteristics

Aggregate addition to acidic red soil significantly improves the water holding capacities (Table 6) of the soil. Addition of 25% of aggregates increased the water holding capacity by 6%. Moreover, aggregates addition significantly improved the hydraulic conductivity (Table 6) of the soil. Addition of 25% of aggregates increased the hydraulic conductivity of the aggregate soil mixture by 10 times. Moreover, aggregate addition up to 25% significantly reduced the bulk density (Table 6) of the red soil from 1.23 gcm^{-3} to 1.05 gcm^{-3},which improves the soil porosity. Addition of aggregates, improves the soil porosity, water holding capacity and soil hydraulic conductivity. Acidic red soil has originally a very low permeability and it leads to heavy erosion due to run off (Hamazaki, 1979). Addition of these stable aggregates can be suggested as an alternative method to minimize the erosion in acidic red soil. Because high Ca content in these aggregates came from CFA can enhance the aggregation of` soil particles together (Jayasinghe & Tokashiki, 2006). Aggregate addition neutralized the acidic soil pH. Aggregate addition percentage of 25% level changed the original pH value of the soil from 4.62 to 6.25, which was nearly a neutral value. Furthermore, electrical conductivity (EC) of soil mixture increases with the addition of aggregates (Table 6). C/N ratios of the soil-aggregate mix also show a significant variance. Application of these aggregates helps to build up C content in the soil as well (Table 6). Aggregate addition of 25 % increased organic carbon from 1.62 to 28.80 gkg^{-1}. Therefore, innovative coal fly ash aggregates addition as a soil amendment improves soil physical and chemical properties of problematic low productive acidic red soil, which automatically improves crop production in the soil.

Characteristics	T1	T2	T3	T4	T5	T6
EC (mSm^{-1})	73.67[a]	55.24[b]	39.64[c]	18.57[d]	10.06[e]	3.27[f]
pH	9.26[a]	8.37[b]	7.86[c]	6.25[d]	5.78[e]	4.62[f]
WHC	0.61[a]	0.59[b]	0.57[b]	0.54[c]	0.53[cd]	0.51[d]
SHC (cms^{-1})	2.20×10^{-2} [a]	6.22×10^{-3} [b]	2.24×10^{-3} [b]	2.94×10^{-4} [c]	2.73×10^{-4} [c]	6.62×10^{-5} [d]
C (gkg^{-1})	110.25[a]	83.18[b]	55.97[c]	28.80[d]	12.56[e]	1.62[f]
N (gkg^{-1})	0.60[a]	0.52[a]	0.54[a]	0.46[a]	0.48[a]	0.47[a]
C/N ratio	200[a]	162[b]	116[c]	65[d]	31[e]	4[f]
BD (gcm^{-3})	0.61[e]	0.87[d]	0.98[c]	1.05[bc]	1.12[b]	1.23[a]

EC: electrical conductivity,
WHC: water holding capacity,
SHC: saturated hydraulic conductivity,
BD: bulk density. (Means followed by the different superscript letter in the same row differed significantly according to Duncan's multiple range test (P=0.05).

Table 6. Physico-chemical characteristics of soil – aggregate amendments

4.1.1.2 Coal fly ash paper waste inorganic binder aggregates (CSA)

CSA (see type E from the Table 1, 2 and 3) were produced by combining coal fly ash and paper waste using an Eirich mixer (R-02M/C27121) with calcium hydroxide and calcium sulfate. 1000 g of coal fly ash and 75 g of paper waste were mixed in the Eirich mixer with 50 g of calcium hydroxide and 50 g of calcium sulfate by adding 350 ml of water to produce CSA. Particle size distribution, physical and chemical properties of CSA were given in Table 1, 2 and 3.

4.1.1.2.1 Scanning Electron Microscopy (SEM)

SEM images of Figure 5-A and B represent detailed micro-morphology of CFA particles. SEM images of CFA particles showed crystalline and amorphous silicate glasses ("stable glasses") of various sizes (Figure 5-A-a) according to 4 typical phases of CFA described by Klose et al ., (2003). CFA consists of many glass-like particles, which are mostly spherical shaped and ranged in particle size from 0.01 to 100µm (Davison et al., 1974). Physically, CFA occurs as very finer particles having an average diameter of <10µm and has low to medium bulk density, high surface area and light texture (Jala & Goyal, 2006). CFA particles are hollow empty spheres called as cenospheres (Figure5-B-c) filled with smaller amorphous particles and crystals (Plerosphers). These tiny CFA particles are easily airborne (Hodgson & Holliday, 1966). Therefore, initial idea of CSA production was to bind tiny airborne CFA particles into fibrous paper waste matrix by starch waste. Figure 5-C and 5-D show the micro-morphological configuration of CSA, where paper waste matrix provides the structural surface to adhere CFA particles. Figure 5-C shows aggregation of CFA particles to paper waste matrix with the assistance of starch. Paper waste matrix increased the surface area of CSA (Figure 5-D). SEM images of SA revealed that SA is a dual composite material having greater surface area with well enmeshed CFA particles in paper waste matrix with the help of starch waste providing porous spaces within the CSA. In a previous study it was reported that CFA showed an increased surface area, capillary action, and nutrient-holding capacity when incorporated to soil (Fisher et al., 1976). Therefore, CSA can be regarded as a material having a higher surface area, which can be utilized to improve the nutrient holding capacity, when incorporated to the soil as a soil amendment.

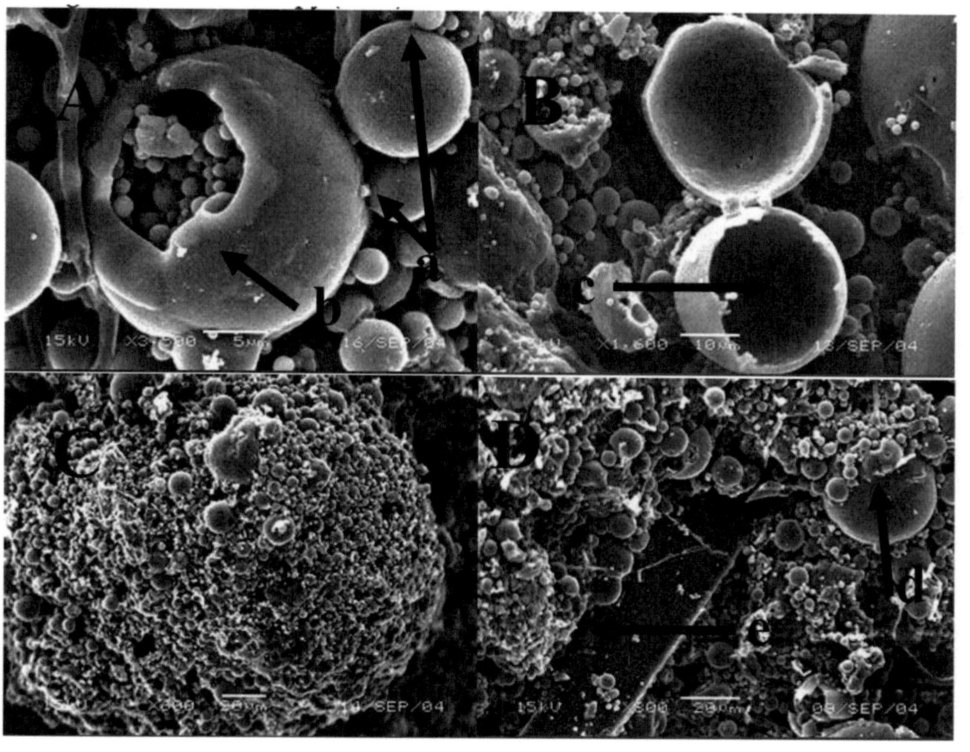

Fig. 5. (A and B) Scanning electron micrographs of CFA particles, showing the micro morphology and varies sizes of coal fly ash particles. (a) stable glasses of various sizes (b and c) cenospheres: (C and D) part of a macro-aggregate (CSA) is produced by CFA and paper waste (d: coal fly ash, e: paper waste).

4.1.1.2.2 Coal fly ash paper waste aggregates as a soil ameliorant to red soil for Komatsuna production

The different aggregate amendment ratios used in the experiment are given in Table 7. Komatsuna (*Brassica rapa* var. Pervidis), also known as Japanese mustard spinach, was grown in a pot experiment to study the influence of CSA amendment in acidic red soil on crop production.

Treatments	Description
T1	Red soil only
T2	CSA :Red soil (1:1) (V/V)
T3	CSA: Red soil (1:5) (V/V)
T4	CSA :Red soil (1:10) (V/V)

CSA: Coal fly ash based synthetic aggregates

Table 7. Different treatments were used under the study.

4.1.1.2.2.1 Effects of CSA amendment on soil physical and chemical properties of the soil

4.1.1.2.2.1.1 Physical properties

CSA addition significantly ($P < 0.05$) decreased the soil bulk density by about 30, 14 and 11% in T2, T3 and T4, respectively. Since, CSA reduced the soil bulk density; it can enhance the porosity and permeability. CSA addition enhanced the hydraulic conductivity of the soil, which may have improved the red soil. Hydraulic conductivity values of treatments T2, T3 and T4 were significantly higher ($P < 0.05$) than that of original red soil (T1) which had a very low hydraulic conductivity of 6.62×10^{-5} cm s^{-1} (Table 8). The addition of CSA increased hydraulic conductivity of T3 and T4 by ten times and T2 by 100 times. CSA addition to red soil also reduced particle density compared with the original soil. Particle densities of the T2, T3 and T4 were reduced by 7.10, 5.28 and 2.26%, respectively in comparison with original soil (T1). Water holding capacity of the T2 (0.59 $kgkg^{-1}$) was increased by 23% compared to the T1 due to incorporation of CSA. This fraction in T3 and T4 were 12.5 and 10%, respectively.

Treatments	Bulk density (gcm^{-3})	Saturated hydraulic conductivity (cms^{-1})	Particle density (gcm^{-3})	Water holding capacity ($kgkg^{-1}$)
T1	1.26^{a}	6.62×10^{-5c}	2.65^{a}	0.48^{c}
T2	0.89^{d}	5.52×10^{-3a}	2.46^{d}	0.59^{a}
T3	1.08^{c}	2.81×10^{-4b}	2.51^{c}	0.54^{b}
T4	1.12^{b}	2.72×10^{-4b}	2.59^{b}	0.53^{b}

(Means followed by the different superscript letter in the same column differed significantly according to Duncan's multiple range test (P=0.05).

Table 8. Physical properties of different treatments used under the study.

The particle size distribution of the different growth media used in the study are shown in Table 9. It showed that red soil had a larger amount of particles < 2 mm. CSA contains 84.02% of particles > 2 mm, while this fraction in red soil (T1) was 44.30%. More over, coal fly ash particles ranging from 0.01 to 100 µm (Page et al. 1979), which can easily become air borne. Production of CSA from CFA significantly increased the particle size diameters (Table 9). CSA production reduced the finer fraction and increased the larger particles, which would reduce handling difficulties. Incorporation of CSA into red soil significantly ($P < 0.05$) increased the fraction > 2 mm by 65.34, 53.34, and 46.36%, in T2, T3 and T4, respectively. Moreover, red soil had a high percentage (29.14 %) of particles < 0.5 mm, while this fraction in CSA, T2, T3 and T4 were 2.23, 10.18, 15.86 and 18.32 %, respectively. CSA as a

Treatments	>5.60mm (Weight %)	5.60-3.35 mm	3.35-2.00 mm	2.00-1.00 mm	1.00-0.50 mm	< 0.50 mm
CSA	26.20^{a}	30.28^{a}	27.54^{a}	8.73^{d}	5.02^{e}	2.23^{e}
T1	6.61^{e}	18.93^{d}	18.76^{c}	14.86^{b}	11.70^{d}	29.14^{a}
T2	19.18^{b}	25.87^{b}	20.29^{b}	11.88^{c}	12.60^{c}	10.18^{d}
T3	14.12^{c}	20.56^{c}	18.66^{c}	14.84^{b}	15.96^{b}	15.86^{c}
T4	9.24^{d}	18.52^{d}	18.60^{c}	17.92^{a}	17.40^{a}	18.32^{b}

CSA: coal fly ash based aggregates, (Means followed by the different superscript letter in the same column differed significantly according to Duncan's multiple range test (P=0.05).

Table 9. Particle size analyses of the CSA, red soil, and CSA-soil amendment mixtures.

soil amendment considerably increased percentage of particles > 2 mm and decreased particles < 0.5 mm. In addition, CSA addition gave a uniform distribution of particles across each particle size class.

4.1.1.2.2.1.2 Chemical properties

The chemical properties of the different treatments are shown in Table 10. The pH values of red soil, coal fly ash and CSA were 5.12, 10.87 and 10.72, respectively. Addition of alkaline CSA to the acidic red soil significantly ($P < 0.05$) decreased the pH, such that T3 and T4 were almost neutral. CSA improved the acidic pH of the red soil to values suitable for plant growth. The EC of the CSA produced was 90.40 mS m^{-1} compared with 4.18 mS m^{-1} in the original red soil. CSA addition to the red soil increased the soil EC; the EC values in T2, T3 and T4 were 60.28, 20.13 and 14.48 mS m^{-1}, respectively. CSA addition to soil significantly increased the Na, K, Mg and Ca concentrations compared to original soil (Table 10). Therefore, CSA addition as a soil amendment not only improves soil physical and chemical properties but also it can improve the soil fertility by supplying nutrients such as Ca, Mg and K. The CEC values of T2, T3 and T4 were significantly ($P < 0.05$) higher than the original red soil, which had CEC of 4.30 $Cmol_ckg^{-1}$ (Table 10). The CEC increments in T2, T3 and T4 were 1.86, 1.76 and 1.62 $cmol_c\ kg^{-1}$, respectively compared to T1. It is evident that CSA addition to the red soil increased the CEC in comparison with the original soil. C content of all treatments with CSA additions increased compared with the original soil (Table 10), due to incorporation of paper waste with a C content of 374.8 g kg^{-1}. The N and P contents of the different treatments were low (0.30-0.50 and 0.03-0.05 g kg^{-1}) and not significantly different between treatments.

Treatments	pH	EC (mSm^{-1})	C (gkg^{-1})	N (gkg^{-1})	P (gkg^{-1})	CEC ($Cmol_ckg^{-1}$)	Na (gkg^{-1})	K (gkg^{-1})	Mg (gkg^{-1})	Ca (gkg^{-1})
T1	5.12^d	4.18^d	1.73^d	0.40^a	0.03^a	4.30^b	0.06^d	0.05^d	0.02^d	0.07^d
T2	8.59^a	60.28^a	18.55^a	0.50^a	0.04^a	6.16^a	0.43^a	0.72^a	0.39^a	11.36^a
T3	7.13^b	20.13^b	7.80^b	0.40^a	0.05^a	6.06^a	0.24^b	0.47^b	0.26^b	3.45^b
T4	6.37^c	14.48^c	4.12^c	0.30^a	0.03^a	5.92^a	0.19^c	0.29^c	0.16^c	1.78^c

(Means followed by the different superscript letter in the same column differed significantly according to Duncan's multiple range test (P=0.05). (n=3). (EC: electrical conductivity, CEC: cation exchange capacity).

Table 10. Chemical properties of different treatments used under the study

Heavy metal concentrations of the different amendment mixtures are given in Table 11. The Cu, Cr, Mn, Zn and Pb concentrations were higher in the amendment mixtures compared with original red soil. Se and Cd were not detected in any treatments, and As was detected only in T2. Heavy metal concentrations of the amendment mixtures were generally well below the maximum pollutant concentration of individual metals for land application suggested by the US Environmental Protection Agency (USEPA, 1993). The average concentrations of heavy metals reported in uncontaminated soils are (all in mg kg^{-1}): As 6, Cr 70, Cu 30, Zn 90, Pb 35, Mn 1000, and Cd 0.35, respectively (Adriano, 2001). The heavy metal concentrations in all amendment mixtures in this experiment were generally below the heavy metal concentrations reported in uncontaminated soils. Kim et al. (1994) reported that heavy metals did not accumulate in a paddy soil following CFA addition at 120 Mg ha^{-1}. Though the concentrations of heavy metals were below uncontaminated soil values and not alarming, there should be routine inspections to ensure that heavy metal concentrations remain within safe limits.

Treatments	Cu ($mgkg^{-1}$)	Cr ($mgkg^{-1}$)	Zn ($mgkg^{-1}$)	Pb ($mgkg^{-1}$)	Cd ($mgkg^{-1}$)	Se ($mgkg^{-1}$)	As ($mgkg^{-1}$)	Mn ($mgkg^{-1}$)
T1	11.7^{b}	5.8^{b}	29.8^{b}	5.6^{b}	ND	ND	ND	20.7^{a}
T2	19.77^{a}	7.1^{a}	44.2^{a}	9.4^{a}	ND	ND	0.1	21.1^{a}
T3	18.54^{a}	6.7^{a}	40.6^{a}	8.3^{a}	ND	ND	ND	20.8^{a}
T4	13.65^{b}	5.9^{b}	35.1^{b}	6.7^{b}	ND	ND	ND	20.3^{a}
USEPA	1500	1200	2800	300	39	36	41	-
Y*	30	70	90	35	0.35	0.4	6	1000

(Means followed by the different superscript letter in the same column differed significantly according to Duncan's multiple range test (P=0.05). (n=3).USEPA =US Environmental Protection Agency standards (1993). Y*: uncontaminated soil (* Adriano, 2001), ND; not detected

Table 11. Heavy metal concentrations of different amendment mixtures.

4.1.1.2.2.2 Influence of CSA as a soil amendment in red soil for Komatsuna cultivation

The growth parameters and the nutrient contents in Komatsuna grown in red soil with different ratios of CSA additions are shown in Table 12. CSA as a soil amendment significantly increased growth and yield parameters of Komatsuna (*Brassica rapa*) compared with the red soil control (T_1). The CSA: soil of 1:5 (T3) and 1:10 (T4) increased plant height and fresh weight yield of Komatsuna about three and 12 times, respectively. The CSA: soil of 1:1 (T2) increased plant height and fresh weight yield by approximately two times and four times, respectively. These yield increases were due to the enhanced physical and chemical properties of the soil from CSA amendment. The CSA addition also enhanced water holding capacity, hydraulic conductivity, CEC and pH compared to original soil, which created a conducive environment to attain higher crop growth and yield parameters. The CSA: soil of 1:5 (T3) and 1:10 (T4) increased soil pH from acidic 5.12, to 7.13 and 6.37, respectively (Table 10). The CECs of T3 and T4 were 40 and 37% higher than the original soil due to CSA incorporation. In a previous study it was reported that, mixed application of CFA and paper factory sludge caused appreciable change in soil physical and chemical properties, increased pH and increased rice (*Oryza sativa*) crop yield (Molliner & Street, 1982). In addition, a mixture of CFA with organic matter is expected to further enhance biological activity in soil (Jala & Goyal, 2006), reduce leaching of major nutrients and beneficial for vegetation (Tripathi et al., 2004).

The nutrient content of plants grown in different substrates is given in Table 12. The N content in shoot tissues from the CSA-amended mixtures was higher than that of the red soil (Table 12). CFA in CSA generally increases plant growth and nutrient uptake (Aitken et al., 1984), and has been shown to supply essential nutrients to crops on nutrient deficient soils and to correct deficiencies of Mg, Ca, K, molybdenum, sulfur and Zn (El-Mogazi et al., 1988). The decreased P content in shoot tissues obtained from red soil is probably due to reduced P availability at a higher soil pH and higher Ca content following CFA amendment (Wong & Wong, 1990). The CSA: soil of 1:1 gave an alkaline pH of 8.59, which did not

Treat ments	Plant height (cm)	Fresh weight (gpot^{-1})	Dry weight (gpot^{-1})	N (%)	P (%)	K (%)	Mg (%)	Ca (%)	Cu (mgkg^{-1})	Mn (mgkg^{-1})	Zn (mgkg^{-1})	Pb (mgkg^{-1})
T1	7.25^{c}	3.84^{c}	0.29^{c}	1.46^{c}	0.56^{a}	2.9^{c}	0.6^{c}	0.5^{b}	3.0^{a}	92.64^{a}	30.4^{d}	0.4^{a}
T2	14.36^{b}	15.19^{b}	2.91^{b}	1.87^{b}	0.26^{d}	3.8^{a}	1.5^{a}	4.5^{a}	3.2^{a}	39.74^{d}	36.7^{a}	0.5^{a}
T3	23.44^{a}	46.95^{a}	8.87^{a}	2.21^{a}	0.32^{c}	3.6^{ab}	1.4^{ab}	4.4^{a}	3.1^{a}	50.12^{c}	35.2^{b}	0.5^{a}
T4	21.03^{a}	43.59^{a}	7.25^{a}	2.18^{a}	0.44^{b}	3.4^{b}	1.2^{b}	4.3^{a}	3.3^{a}	54.16^{b}	34.1^{c}	0.4^{a}

(Means followed by the different superscript letter in the same column differed significantly according to Duncan's multiple range test (P=0.05).(n=3).

Table 12. The growth parameters and the nutrient content in the Komatsuna plants.

improve growth and yield parameters of Komatsuna, which needs a pH of 6.0–7.5 for healthy growth. It is likely that the high pH (8.59), high EC (60.28 mS m^{-1}) and its related nutrient bioavailability accounted for the reduced growth and yield parameters of Komatsuna in the T2 treatment compared with T3 and T4. High pH of substrate can sharply decrease availabilities of P, iron (Fe) and Mn (Peterson, 1982). The lowest shoot concentration of Mn was given by plants grown in T2 treatment (Table 12). Similar results were found in a previous study conducted using CFA as an amendment to container substrate for *Spathiphyllum* production by Chen & Li (2006). The growth and yield parameters of Komatsuna grown in red soil were significantly lower than the CSA-amended treatments. Red soil had acidic pH of 5.12, which can decrease plant availability of Ca and Mg but increase solubility of micronutrients. Shoot concentrations of Ca and Mg were lower but Mn was higher in plants grown red soil (T1) than those grown in CSA addition treatments (Table 12). Additionally, low pH was reported to directly affect the permeability of root cell membranes and leakage of various ions from roots (Yan et al., 1992). Moreover, the K, Mg and Ca concentrations of shoots in the CSA mixtures significantly increased because CSA was enriched with these elements. Both shoot K, Ca and Mg contents were all above the deficiency limits of 0.7-1.5 % (Chapman 1966), 0.14 % (Loneragen & Snowball, 1969) and 0.06 % (Chapman, 1966), respectively. The highest Mn concentration was reported in the shoots from the red soil. Nevertheless, the Mn concentrations were much higher than the diagnostic deficiency level of 20 mg kg^{-1} (Chapman 1966). Zn concentrations in the CSA mixtures were higher than in red soil but well below the toxicity limit of 150 mg kg^{-1} (Elseewi et al., 1980). Se and Cd were not detected in any tissues. Therefore heavy metal content in plant tissue was below the toxicity limits. CSA as a soil amendment can be suggested as a good practice for Komatsuna production in this low-productive acidic red soil, due to enhanced soil physical and chemical properties. The CSA: soil of 1:5 and 1:10 gave the maximum growth and yield parameters of Komatsuna.

4.1.1.3 Other types of coal fly ash based aggregates as a soil ameliorant

CFA based aggregates can be developed by including oil palm waste and coco fiber and also can be used as a soil amendment to improve low productive red soil to enhance crop production. Since CFA aggregates did not contain high N content, aggeragets can be developed by adding N fertilizer source to the aggeragets and those aggregates can be used as a fertilizer and a soil ameliorant for crop production. One of our experiments (Jayasinghe et al., 2009b) showed that these kind of N added aggeragets improved physical and chemical properties of low productive acidic soil and improved the growth and yield parameters of Komatsuna in red soil compared to original soil. In addition, homogenous synthetic aggeragets also can be produced with CFA, paper waste and organic or inorganic binder and can be used as a soil ameliorant for low productive acidic soils. Moreover, CFA, paper waste and starch binder pellet aggregates as a soil ameliorant for acidic red soil and problematic grey soil improved the respective chemical and physical properties of the soil and crop growth and yield of *Brassica campestris.* In conclusion, we can emphasis that different types of aggregates produced from different waste materials can be effectively utilized as a soil ameliorant to enhance the crop production.

4.1.1.4 Synthetic red soil aggregates

4.1.1.4.1 Synthetic aggregates (SA) production

SA was produced by combining red soil and paper waste in an Eirich mixer (R-02M/C27121) with starch as the binder. First of all 1000g of red soil and 125g of paper wastes were mixed well in the pan of the Eirich mixer for 2 minutes. Subsequently, 225 mL of prepared starch paste was added to the above mixture and mixed well for another 2 minutes to produce SA (30 g of starch wastes was added to 200 mL hot water in 50°C and heated to 80°C in order to obtain sticky paste as the aggregate binder).

4.1.1.4.2. Effect of synthetic soil aggregates as a soil amendment to enhance properties of problematic grey ("Jahgaru") soils in Okinawa, Japan.

Grey soils ("Jahgaru") in Okinawa Japan spread over 20% of the total land area showed low infiltration, strong stickiness and plasticity and alkalinity (National Institute of Agro Environmental Sciences, 1996). It also exhibits a poorly developed soil structure and poor air and water permeability characteristics (Okinawa Prefecture Agricultural Experiment Station, 1999). Crop production on this grey soil is challenged due to possible disasters (i.e. drainage problems, poor permeability), which can be resulted due to poor properties of the soil. Moreover, soil structure has been found to be of paramount importance in soil productivity and is becoming limiting factor of the crop yield (Allison, 1973).Therefore, SA was utilized as a soil ameliorant to improve the problematic properties of grey soil to enhance ornamental plant production. Therefore, French marigold (***Tagetes patula***) was selected as the ornamental plant in this study. The objectives of the present study were to study the characteristics of the grey soil amended with SA and to study the influence of the SA addition as a soil ameliorant on the growth parameters of French marigold (***Tagetes patula***). Experiments were conducted to study the impact of SA as a soil amendment to improve poor properties of grey soil. Widely using French marigold (***Tagetes patula***), which is a popular ornamental plant in Japan was grown in the experiment. All 8 treatments are shown in Table 13. Grey soil was amended with SA at the rates of 10%, 20%, 30%, 40%, and 50% respectively.

4.1.1.4.2.1.Effect of aggregate addition on particle size distribution and Mean weight diameter (MWD) of soil

Particle size distribution and mean weight diameter (MWD) of different treatments used in the experiment are given in Table 14. Particle size distribution of a substrate is important because it determines pore space, bulk density, air and water holding capacities (Raviv et al., 1986). The mean particle size distribution of treatments showed that the fraction < 0.25 mm was the most abundant fraction in T1 (29.28 %) and T8 (22.47 %). More over, the particle percentages > 3.35 mm are the lowest in T1 (grey soil) and T2 (red soil) treatments.

Treatments	Description
T1	Grey soil only
T2	10% SA addition
T3	20% SA addition
T4	30% SA addition
T5	40% SA addition
T6	50% SA addition
T7	SA only
T8	Red soil only

SA: synthetic aggregates

Table 13. Different treatments utilized under the study

Treatments	>5.60 mm	5.60-3.35 mm	3.35-2.00 mm	2.00-1.00 mm	1.00-0.50 mm	0.50-0.25 mm	0.25-0.00 mm	MWD (mm)
T1	0.00	5.08h	15.02f	14.98c	17.47b	18.17b	29.28a	1.090h
T2	0.34f	11.72f	16.30e	17.58b	15.17c	16.57c	22.32c	1.452f
T3	1.25e	13.75e	16.26e	17.54b	16.90b	14.97d	19.33d	1.618e
T4	2.19d	17.49d	19.81d	15.22c	14.59c	13.66e	17.04e	1.894d
T5	4.15c	19.90c	23.42c	17.65b	15.02c	10.50f	9.36f	2.270c
T6	5.36b	22.80b	24.29b	16.97b	14.11d	8.83g	7.64g	2.491b
T7	9.77a	26.82a	30.98a	18.15a	5.91e	4.28h	4.09h	3.129a
T8	0.00	8.28g	11.23g	18.85a	20.15a	19.02a	22.47b	1.204g

(Means followed by the different letter in the same column differed significantly according to Duncan's multiple range test (P=0.05).Values are mean (n=3)).

Table 14. Particle size distribution and Mean weight Diameter (MWD) of the different treatments used in the experiment

The highest particle percentage >3.35 mm was given by T7 (SA only). Production of SA from red soil with paper waste significantly (P<0.05) increased the particle size diameters compared to red soil only. In a previous study conducted by Jayasinghe etal., (2008) to produce SA from red soil and coal fly ash, gave increased particle size diameters compared to original red soil. Dominance of finer particles in a substrate clogs pores, increase non-plant available water holding capacity and decrease air filled porosity (Spiers & Fietje,2000), while dominance of larger particles in a substrate increase aeration and decrease water retention (Benito et al., 2006). Accordingly, red soil and grey soil had higher percentage of finer particles, which led to decrease air filled porosity of the medium. Addition of higher percentage of SA as a soil amendment significantly (P<0.05) decreased the finer particles < 0.25 mm and significantly (P<0.05) increased the larger particles. Moreover, addition of higher SA percentages significantly increased the MWD compared to original soil (Table 14). MWD of grey soil (1.090mm) had increased to 1.452, 1.618, 1.894, 2.270, 2.491 mm at SA amendment of 10%, 20%, 30%, 40%, and 50% of SA, respectively. It is evident that the amelioration of the grey soil with SA had significantly (P<0.05) increased the MWD of original grey soils.

4.1.1.4.2.2 Effect of aggregate addition on Water stable aggregates (WSA), organic matter content and C content

The percentage of WSA > 0.25 mm, organic matter content and the C content of different treatments are given in the Figure 6. It is evident that WSA percentage is significantly ($P<0.05$) increased with the addition of SA to grey soil. WSA varied from 40.72 % to 86.63 % and the lowest WSA given by grey soil with no SA amendment (T1) and the highest was given by SA only (T7). Aggregate stability, a measure of the soil's resistance to externally imposed disruptive forces, was increased with increasing SA amendment percentages. It has been shown that the addition of organic matter improved soil properties such as aggregation, water holding capacity, hydraulic conductivity, bulk density, the degree of compaction, fertility, and resistance to water and wind erosion (Franzluebbers, 2002). Generally, crop residues, turfs, paper wastes, manures, forest under story leaf falls, and compost from organic wastes have been used to increase soil organic matter content and accordingly to improve soil physical properties in crop lands (Stratton et. al., 1995).

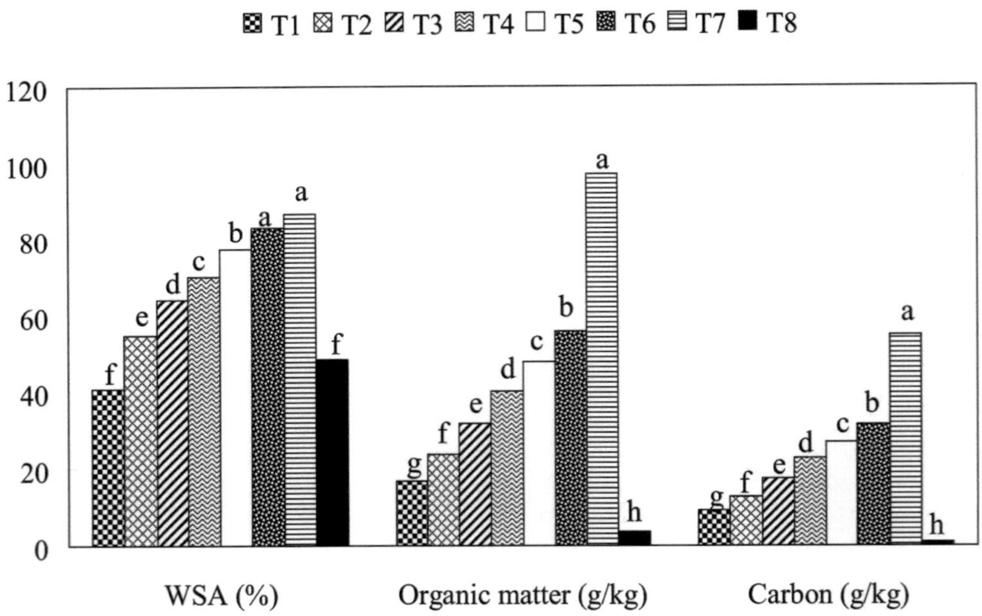

Fig. 6. Water stable aggregates (WSA), organic matter content and the carbon content of the different treatments studied under the study.(Different letters on the top of the bars differed significantly according to Duncan's multiple range test, n=3).

Aggregation is maintained by the presence of organic matter (Lynch & Bragg, 1985) and therefore changes in organic matter content can lead to changes in aggregation (Dexter, 1988). SA addition increased the soil organic matter (Figure 6) due to incorporation of paper waste and starch waste to develop SA. Paper waste in the SA is also a rich source of C (Rasp & Koch, 1992) and improves soil organic matter contents, water holding capacity, soil structure and bulk density (Simard et al., 1998). Highest organic matter content is reported in SA only (97.15 gkg^{-1}). The lowest percentage of organic matter was given by red soil (3.78 gkg^{-1}) and the second lowest organic matter content was given by the grey soil (16.74 gkg^{-1}). Addition of SA to grey soil significantly ($P<0.05$) increased the organic matter content. Moreover, high correlation between aggregate stability and soil organic matter has been reported by Chaney & Swift (1984). More over SA addition significantly enhanced the soil C content (Figure 6). The C amounts of red and grey soils were 1.16 and 9.42 gkg^{-1}, respectively, which were the two lowest values reported among all other treatments. Original red soil and the grey soil showed very low amount of C. T7 gave the highest amount of C (54.72 gkg^{-1}) content due to the paper waste and starch waste. The C content of the paper waste and starch waste were 374.8 gkg^{-1} and 312.5 gkg^{-1}, respectively. Therefore, SA can be considered as a significant source of C. Several previous studies revealed that a trend of positive relationship between C content of the soil and the aggregate stability and MWD (Adesodun et al., 2005). Soil aggregation, which is important to crop establishment, water infiltration and resistance to erosion and compaction, is also influenced by soil C content (Wright & hons, 2005).

4.1.1.4.2.3. Effect of aggregate addition on physical properties of Soil-SA amendment mixtures

Physical properties of different amendment mixtures are given by Table 15. The bulk density of the treatments has been decreased with the increasing percentages of SA amelioration. SA addition significantly ($P<0.05$) decreased the soil bulk density by about 5.17 %, 9.48 %, 13.79%, 17.24% and 21.55% in SA addition treatments of T2, T3, T4, T5 and T6, respectively. Bulk density depends on soil structure and is an indicator of soil compaction, aeration and development of roots especially in soils with high clay contents. The bulk density of the grey soil was 1.16 gcm^{-3}.The lowest bulk density (0.85gcm^{-3}) was given by SA only treatment. The organic matter content in red soil, SA and grey soil were 3.78, 16.74 and 97.15 gkg^{-1}, respectively. Bulk density decreased with the increased amount of soil organic matter (Nyakatawa et al., 2001). Moreover, red soil showed the highest bulk density (1.21 gcm^{-3}) and it was reduced to 0.85 gcm^{-3} by 29.95% after producing SA with paper waste. Porosity values were increased by 3.41 %, 5.82 %, 8.63 %, 11.04 % and 13.68 % in SA addition treatments of T2, T3, T4, T5 and T6, respectively. Highest porosity (63.68 %) was given by the T7 treatment. The increased porosity is especially important to crop development since it may have a direct effect on soil aeration and can enhance root growth (Sugiyanto et al., 1986).SA addition significantly ($P<0.05$) increased the water holding capacity of the soil compared to grey soil only (T1). Water holding capacity was increased by 8.51 %, 10.64%, 19.15 %, 23.40 % and 8.51 %, in T2, T3, T4, T5 and T6, respectively compared to T1. Red soil (T8) gave the lowest water holding capacity (0.43 $kgkg^{-1}$) and it was increased to 0.50 kg kg^{-1} in SA only (T7) by 16.28 % compared to red soil after converting into SA with paper waste. It can be suggested that, water holding capacity increases are due to the improved soil microspores, which resulted after SA addition, subsequently responsible for the holding of water within the soil. Hydraulic conductivity values of SA addition treatments showed significant differences ($P<0.05$) compared to T1 control except T2 (Table 15). Hydraulic conductivity values were increased by 10 times in SA addition treatments of T3, T4, T5 and T6 compared to original grey soil (T1).

Treatments	BD (gcm^{-3})	PD (gcm^{-3})	Porosity (%)	WHC ($kgkg^{-1}$)	Permeability (cms^{-1})
T1	1.16ab	2.54a	54.33g	0.47c	6.671×10^{-5}a
T2	1.10b	2.51ab	56.18f	0.51b	7.471×10^{-5}a
T3	1.05c	2.47b	57.49e	0.52b	1.031×10^{-4}b
T4	1.00cd	2.44bc	59.02d	0.56a	1.334×10^{-4}b
T5	0.96d	2.42c	60.33c	0.58a	1.031×10^{-4}b
T6	0.91e	2.38cd	61.76b	0.51b	1.867×10^{-4}b
T7	0.85f	2.34d	63.68a	0.50b	1.601×10^{-3}c
T8	1.21a	2.58a	53.10h	0.43d	4.916×10^{-5}a

(Means followed by the different letter in the same column differed significantly according to Duncan's multiple range test (P=0.05).Values are mean (n=3)).

Table 15. Physical properties of the treatments used in the study

4.1.1.4.2.4 Aggregate addition on chemical properties of Soil-SA amendment mixtures

Chemical properties of the different treatments are shown in Table 16. The pH of grey (T1) and red soil (T2) were 8.36 and 4.46, respectively. It is evident that grey soil is an alkaline soil and the red soil is an acidic soil. Produced SA from red soil also showed acidic pH of 4.42. It is evident that mixing of the acidic SA to grey soil decreased the alkaline pH values. For an example 30% SA addition to grey soil gave a pH value of 7.72. EC of grey soil showed significantly ($P<0.05$) higher values compared to red soil and other treatments. This may be due to high concentrations of cations in grey soil (Table 16). The highest concentrations of Ca, Mg, K and Na concentrations were given by grey soil while the lowest was given by red soil. The N content of the grey soil and the SA amendment did not show any significant differences.

Treatments	pH	EC	Ca (gkg^{-1})	Mg (gkg^{-1})	K (gkg^{-1})	Na (gkg^{-1})	N (gkg^{-1})
T1	8.36a	69.3a	11.45a	1.28a	0.04a	0.10a	0.74a
T2	8.28a	67.8b	10.34b	1.19b	0.04a	0.10a	0.69a
T3	8.10b	65.1b	9.25c	1.11c	0.04a	0.09a	0.65a
T4	7.72c	62.7c	8.27d	1.03d	0.03a	0.08b	0.62a
T5	7.51c	60.2c	7.23e	0.96e	0.03a	0.08b	0.58a
T6	7.42d	58.8c	6.19f	0.88f	0.03a	0.07b	0.55a
T7	4.42e	8.82d	0.97g	0.51g	0.01b	0.04c	0.40b
T8	4.46e	5.36e	0.94g	0.54g	0.01b	0.04c	0.42b

(Means followed by the different letter in the same column differed significantly according to Duncan's multiple range test (P=0.05). (n=3)).

Table 16. Chemical properties of the treatments studied under the experiment

4.1.1.4.2.5 Influence of SA addition on plant growth parameters

It is evident that the plant growth parameters of French marigold were significantly ($P<0.05$) increased with the addition of SA as a soil ameliorant to grey soil (Table17). The lowest growth parameters are reported in French marigold grown in the grey and red soil, while the highest growth and yield parameters were given by 30 % (T4) SA addition. The plant height, number of flowers per plant, shoot fresh weight, shoot dry weight, root length, root fresh weight and root dry weight in T4 were increased by 1.5, 2.9, 3.5, 4.7, 3.4, 4.3 and 9 times, respectively compared to T1. These increased growth parameters were because of the enhanced particle size distribution, aggregate stability, soil organic matter, soil C, bulk density, soil porosity, water holding capacity, hydraulic conductivity and pH due to the SA amelioration compared to original grey soil. An increase in pore size and the continuity of pore space eases root penetration and flow of water and gases, directly related to plant growth (Tejada & Gonzalez, 2003).The lowest growth parameters of French marigold in grey soil and red soil may be due to its poor soil structure. Improvement in soil aggregation by aggregate addition positively affects germination of seeds and the growth and development of plant roots and shoots (Van Noordwijk et al., 1993). The highest root length (190.33 mm) was reported in the SA only treatment (T7) due to the increased porosity. The root length of French marigold has been increasing with increasing percentage of SA addition. This may be due to the improved soil porosity of the soil after SA addition. More over, growth reduction in red soil (T8) and SA only (T7) may be due to acidic pH. The acidic pH of the red soil and the SA were 4.46 and 4.42, respectively. The pH values of 30% and 40% SA additions to grey soils were 7.72 and 7.52 compared to grey soil pH of 8.36.

Treatments	Plant height (cm)	Number of flowers per plant	Shoot fresh weight (gplant^{-1})	Shoot dry weight (gplant^{-1})	Root length (mm)	Root fresh weight (gplant^{-1})	Root dry weight (gplant^{-1})
T1	14.23f	6.33f	5.89e	0.72e	53.30f	0.45g	0.03e
T2	16.27d	8.66d	7.81d	0.96d	90.00d	0.63f	0.05e
T3	18.21b	15.66b	14.22b	2.30b	120.33c	1.35d	0.11d
T4	21.42a	18.66a	20.74a	3.35a	180.00b	1.92a	0.27a
T5	20.66a	17.66a	20.24a	3.26a	178.66b	1.85a	0.25a
T6	17.13c	14.00b	11.64c	2.38b	182.00b	1.41c	0.16c
T7	16.33d	11.00c	10.63c	1.32c	190.33a	1.23e	0.11d
T8	15.66e	7.33e	6.29d	0.85f	84.66e	0.48g	0.04e

(Means followed by the different letter in the same column differed significantly according to Duncan's multiple range test (P=0.05). (n=3)).

Table 17. Effects of SA amendment on the growth of French marigold

The mineral element concentrations in French marigold shoots are given in Table 18. It is evident that shoot concentrations of N in SA addition treatments have been increasing compared with the T1 control but again decreasing in T7 and T8 treatments. Moreover, plants grown in grey soil and SA amendments gave higher K, Mg and Ca concentrations in plant shoots compared to plants grown in red soil and SA. It is evident that shoot concentrations of K, Ca and Mg were low but Mn concentration was significantly ($P<0.05$) higher in plants grown in red soil (T8) and the SA only (T7) compared with other treatments. This may be due to the acidic pH values T7 and T8 treatments, which can decrease bio availability of Ca and Mg but increase solubility of micronutrients. It is likely that the low pH and its related nutrient bioavailability accounted for the reduced growth of French marigold grown in red soil and SA only treatments. In addition, the lowest iron (Fe) concentration was reported in the plants grown in grey soil (T1). This may be due to the high pH of the grey soil compared to other treatments. It has been found that high pH of the substrate sharply decrease the availability of Fe in the substrate (Peterson, 1982). Addition of 30% to 40% of SA to problematic grey soil can be recommended as a better practice for French marigold production, which gave the best growth parameters compared to original grey soil.

Treatments	N (gkg^{-1})	K (gkg^{-1})	Mg (gkg^{-1})	Ca (gkg^{-1})	Fe ($mgkg^{-1}$)	Cu ($mgkg^{-1}$)	Mn ($mgkg^{-1}$)	Zn ($mgkg^{-1}$)
T1	29.21d	33.71a	12.18a	42.49a	44.78c	3.49b	62.01c	33.10b
T2	30.86c	33.65a	12.22a	42.44a	45.91c	3.54b	61.98c	33.21b
T3	31.24b	33.64a	12.24a	42.27a	46.77c	3.56b	63.41b	33.15b
T4	33.12a	32.58a	12.37a	42.30a	46.82b	4.07a	64.16b	35.11a
T5	33.25a	32.49a	12.28a	42.36a	47.28b	4.19a	65.55b	35.23a
T6	33.18a	32.31a	12.21a	42.24a	47.12b	4.10a	68.75b	35.47a
T7	24.16e	22.32b	8.38b	6.72b	79.25a	3.24c	109.56a	32.12c
T8	24.28e	22.45b	8.45b	6.77b	78.54a	3.22c	110.72a	32.09c

(Means followed by the different letter in the same column differed significantly according to Duncan's multiple range test (P=0.05).Values are mean (n=3)).

Table 18. Element concentrations in plant shoots

4.2 Synthetic aggregates as a potting media

4.2.1 Coal fly-ash-based synthetic aggregates as alternative container substrates for ornamentals.

Experiment was conducted using CFA paper waste aggregates and oil palm waste as a potting media for ornamental French marigold (***Tagetes patula***) production. Utilized container substrates under the experiment are given in table 19. Zeolite was the standard substrate.

Substrate	Formulation
T1	CSA (100%)
T2	CSA 1 :Palm waste 5 (V/V)
T3	CSA1 :Palm waste 10 (V/V)
T4	Palm waste (100%)
T5	Zeolite

CSA: coal fly ash based synthetic aggregates

Table 19. Composition of container substrates used in the experiment

4.2.1.1 Physical properties of the container substrates

The particle size distribution of different substrates utilized in this study is given by the Table 20. The mean distribution showed that the fraction greater than 3.5 mm was the most abundant fraction in CSA and zeolite substrates. 55.80 % of the CSA particles were greater than 3.5 mm while this fraction in zeolite, oil palm waste, T2 and T3 were 80.52%, 8.44 %, 34.68% and 21.76%, respectively. The best substrate is that with medium to coarse texture, equivalent to a particle size distribution between 0.25 and 2.00 mm, as the optimal range for a plant growth medium that allow retention of enough readily available water together with adequate air content (Abad et al., 1993; Benito et al., 2006). T1 and T5 gave less than 15% of particles between 0.25 and 2.00 mm whereas this fraction in T2, T3 and T4 were 45%, 62% and 80%, respectively. CSA (T1) and Zeolite (T5) did not give sufficient particle percentages in the optimal range (0.25 and 2.00 mm) for plant growth whereas oil palm waste gave the highest fraction in this range. T2 and T3 had higher percentages of particles between 2.00 and 0.25 mm compared with T1 and T5. Moreover, T2 and T3 had a uniform distribution of particles in every particle size diameter class because of mixing comparatively larger particles of CSA and smaller particles of palm waste.

Substrate	>5.6mm (Weight %)	5.6-3.35 mm	3.35-2.00 mm	2.00-1.00 mm	1.00-0.50 mm	0.50-0.25 mm	<0.25 mm
T1	25.61^{a}	30.19^{b}	28.32^{a}	8.54^{d}	5.20^{d}	1.08^{e}	1.06^{d}
T2	14.12^{b}	20.56^{c}	18.76^{b}	13.84^{c}	16.96^{c}	14.05^{c}	1.71^{c}
T3	8.24^{c}	13.52^{d}	14.60^{c}	23.92^{b}	19.40^{b}	18.44^{b}	1.88^{b}
T4	0	8.44^{e}	9.24^{e}	29.24^{a}	22.48^{a}	28.37^{a}	2.23^{a}
T5	0	80.52^{a}	13.09^{d}	2.46^{e}	1.14^{e}	1.62^{d}	1.17^{d}

(Means followed by the different superscript letter in the same column differed significantly according to Duncan's multiple range test (P=0.05).Values are mean (n=3))

Table 20. Particle size analysis of the container substrates

Table 21 shows the main physical properties of the different container substrates used in the study. Bulk density values of different substrates were significantly different (P<0.05). Abad et al., (2001) defined the bulk density requirement of an ideal substrate should be less than 0.40 gcm^{-3}. CSA and the zeolite media exceeded these limits. However T2, T3, and T4 substrates were within the ideal range. Mixing CSA with oil palm waste at the ratio of 1:5 and 1:10 decreased the bulk density by 51% and 55% compared to CSA (T1) and 63% and 65% compared to zeolite (T5), respectively. Particle density values of T2, T3, and T4 are in the range of established particle density limit (1.4-2.0 gcm^{-3}) described by Abad et al., (2001). Air space values for an ideal substrate should be within 20-30% according to De Boot & Verdonck, (1972). Zeolite showed highest air space while oil palm waste showed lowest

Substrate	BD (gcm^{-3})	PD (gcm^{-3})	Air space (%)	Total pore space (%)	Total water holding capacity (mLL^{-1})
T1	0.56^{b}	2.48^{b}	25.82^{b}	77.42^{c}	516^{d}
T2	0.27^{c}	1.81^{c}	22.08^{c}	85.08^{b}	630^{c}
T3	0.25^{d}	1.76^{d}	20.59^{d}	85.79^{b}	652^{b}
T4	0.22^{e}	1.70^{e}	19.36^{e}	87.06^{a}	677^{a}
T5	0.72^{a}	2.61^{a}	31.71^{a}	72.41^{d}	407^{e}
IS^{X}	<0.40	1.4-2.0	20-30	>85	600-1000

IS^{X}: Ideal Substrate according to De Boodt and Verdonck, (1972) and Abad et al., (2001); BD: bulk density, PD: particle density:(Means followed by the different superscript letter in the same column differed significantly according to Duncan's multiple range test (P=0.05).Values are mean (n=3)).

Table 21. Physical properties of the container substrates

value compared to the established ideal range. High air space means that water should be applied frequently, and in small amounts to avoid leaching (Benito et al., 2006). The ideal total pore space should exceed 85% (De Boodt & Verdonck, 1972). T2, T3, and T4 growth media exceeded this ideal limit while other media were below than the ideal value. Ideal total water holding capacity of an ideal substrate should be in the range of 600-1000 mLL^{-1} according to De Boodt and Verdonck, (1972). T2, T3, and T4 substrates were in the ideal substrate range whereas T1 and T5 were not within the ideal range. Physical properties such as bulk density, particle density, air space, total pore space and total water holding capacity of the T2 and T3 substrates were in the ideal substrate range compared to T1 and T5. Moreover, the physical properties of the T4 substrate were in the ideal range except the air space.

4.2.1.2 Chemical properties of the container substrates

The CSA (T1) showed an alkaline pH value of 9.82 (Table 22). Oil palm waste and Zeolite showed pH values of 4.34 and 8.77, respectively. Mixing ratio of CSA with oil palm waste at the ratio of 1:5 and 1:10 gave pH values of 7.21 and 6.18. Established pH limits for an ideal substrate are 5.3-6.5 (Abad et al., 2001). T3 was the only substrate lies within the ideal substrate range with respect to pH values. The EC values showed significant differences ($P<0.05$) among all substrates. Established EC limits for an ideal substrate are 50 mSm^{-1} (Abad et al., 2001). The T2, T3, and T4 substrates were in the range of established ideal substrate range while EC values of T1 and T5 were not in the ideal range. C contents of all container substrates were significantly differed ($P<0.05$) with each other. Nitrogen content of all container substrates was very low and varied between 0.4 and 5.2 gkg^{-1}. The carbon and nitrogen (C/N) ratio, an indicator of organic matter origin, was different for all the substrates. The CSA and the zeolite showed the highest C/N ratios because of the low nitrogen contents. High C/N ratios could be cause immobilization of soluble nitrogen when those substrates were used as a growing medium for containerized plant production. Established range of C/N ratio for ideal substrate is 20-40 (De Boodt & Verdonck, 1972). T2, T3 and T4 substrates were closest to the ideal value. CSA and the zeolite media showed significantly ($P<0.05$) higher concentrations of cations (Table 22) compared to other substrates. Oil palm waste showed the lowest concentration of the cation compared to all substrates studied under this study. In addition, substrates having a mixture of CSA and oil palm waste (T2 and T3) reduced the cation concentration compared to CSA (T1) and zeolite (T5). CSA showed the highest concentration of heavy metals compared with other substrates. Cd and Se were not detected in all substrate samples. The standard maximum pollutant concentrations of individual metals for land application suggested by the US Environmental protection Agency (USEPA, 1993) are 41 $mgkg^{-1}$As, 1200 $mgkg^{-1}$Cr, 1500 $mgkg^{-1}$Cu, 2800 $mgkg^{-1}$Zn, 300 $mgkg^{-1}$Pb and 39 $mgkg^{-1}$ Cd, respectively. It is evident that, heavy metal concentrations in all substrates used in this experiment were generally below the maximum pollutant concentration of individual metals for land application suggested by the US Environmental protection Agency (USEPA, 1993).

4.2.1.3 Effect of different substrates on the growth and yield parameters of French marigold.

The shoot fresh weight, shoot dry weight, root fresh weight, root dry weight, plant height and number of flowers per Marigold plant grown in different container substrates after 3 months of growing period are shown in Table 23. There were significant differences ($P<0.05$) in growth and yield parameters of Marigold grown in different growth substrates. The lowest shoot fresh weight, shoot dry weight, root fresh weight, root dry weight and plant

Substrate	pH	EC (mSm^{-1})	C (gkg^{-1})	N (gkg^{-1})	C/N ratio	Na (gkg^{-1})	K (gkg^{-1})	Mg (gkg^{-1})	Ca (gkg^{-1})	As ($mgkg^{-1}$)	Cr ($mgkg^{-1}$)	Cu ($mgkg^{-1}$)	Mn ($mgkg^{-1}$)	Zn ($mgkg^{-1}$)	Pb ($mgkg^{-1}$)
T1	9.82^a	96.1^a	120.82^d	0.71^d	172.60^a	0.87^a	1.51^a	0.72^a	3.34^b	0.21^a	7.62^a	18.47^a	15.82^a	34.63^a	7.56^a
T2	7.21^c	48.8^c	179.11^c	3.58^c	49.75^c	0.34^c	0.72^c	0.37^c	1.41^c	0.12^a	5.13^b	10.82^b	12.61^c	25.12^b	6.72^b
T3	6.18^d	42.4^d	190.24^b	4.22^b	45.29^d	0.25^d	0.55^d	0.29^d	1.03^d	0.11^a	3.59^c	8.68^c	10.67^d	11.61^d	5.89^c
T4	4.34^e	32.3^e	210.80^a	5.17^a	40.54^e	0.10^e	0.37^e	0.21^e	0.46^e	0.12^a	0.28^e	2.02^d	2.81^e	1.17^e	0.41^d
T5	8.77^b	59.8^b	28.27^e	0.38^e	70.68^b	0.56^b	0.91^b	0.45^b	3.78^a	0.22^a	1.41^d	11.08^b	13.78^b	21.72^c	6.08^b

EC: electrical conductivity (Means followed by the different superscript letter in the same column differed significantly according to Duncan's multiple range test (P=0.05). Values are mean (n=3)).

Table 22. Chemical properties of the container substrates

Substrate	Shoot Fresh weight ($gplant^{-1}$)	Shoot dry weight ($gplant^{-1}$)	Root fresh weight ($gplant^{-1}$)	Root dry weight ($gplant^{-1}$)	Plant height (cm)	No of flowers per plant
T1	11.68^e	1.59^e	4.55^e	0.65^e	9.60^e	0
T2	141.59^b	19.54^b	49.25^b	7.39^b	31.33^b	56.00^b
T3	165.17^a	23.87^a	56.34^a	8.90^a	34.00^a	64.00^a
T4	101.37^d	11.00^d	33.88^d	3.31^d	25.33^d	38.33^d
T5	109.37^c	12.37^c	36.36^c	3.56^c	28.66^c	39.66^c

(Means followed by the different superscript letter in the same column differed significantly according to Duncan's multiple range test (P=0.05).Values are mean (n=3)).

Table 23. Effects of container substrates on the growth of French marigold

height were given by plant grown in T1 substrate. French marigold requires a substrate pH of 5.5-6.8 for healthy growth. The reduced growth and yield parameters of the French Marigold grown in T1 substrate are due to the high substrate pH (9.82) and EC (96.1 mSm^{-1}) and its related nutrient bioavailability. Substrate T3 was most sufficient to promote the growth of French marigold because of its optimal chemical and physical properties (e.g.pH, EC, bulk density, total pore space, water holding capacity, etc.). The growth and yield parameters of French marigold grown in the zeolite (T5) were significantly ($P<0.05$) lower than that of T2 and T3 substrates but significantly higher than that of oil palm waste (T4) and CSA (T1). Shoot concentrations of Ca and Mg were low but Mn concentration was significantly higher in plants grown in oil palm waste (T4) compared with other treatments. This may be due to the acidic pH value (4.34) of oil palm waste, which can decrease bio availability of Ca and Mg but increase solubility of micronutrients. pH and its related nutrient bioavailability accounted for the reduced growth of French marigold grown in the oil palm waste. Mixing CSA with the oil palm waste at the ratio of 1:10 (T3), which is an ideal substrate, can be suggested as an alternative container substrate for French marigold production compared with zeolite. In addition, production of CSA using waste CFA with paper waste and mixing them with oil palm waste as a container substrate can be suggested as an alternative waste management practice.

4.2.2 Utilization of synthetic soil aggregates as a containerized growth medium component to substitute peat in the ornamental plant production

An investigation was under taken to study the characteristics and utilization of synthetic soil aggregates (SA) formed by low productive acidic soil with paper and starch waste for production of French marigold (*Tagetes patula*) as a partial peat substitution in growing substrate. Five different growth substrates utilized in this study were peat only, peat 75%: SA 25%, peat 50%: SA 50%, peat 25%: SA 75% and SA only. Peat 75%: SA 25% enhanced substrate physical and chemical properties into the established ideal substrate range (Jayasinghe et al., 2009 e). Plant height, numbers of flowers, fresh shoot weight, dry shoot weight, root length, fresh root weight and dry root weight of French marigold grown in the substrate of peat 75%: SA 25% increased by 13.28, 23.07, 28.51, 27.41, 6.66, 68.33 and 7.40%, respectively compared with peat substrate. Peat 50%:SA50% gave similar growth parameters to peat only substrate. Nitrogen (N) content of plants grown in peat 75%:SA25% was higher than peat substrate. The normal ranges of Cu, Fe, Mn and Zn in plants were 3-20, 30-300, 15-150 and 15-150 mg kg^{-1}, respectively (Adriano 2001). The critical toxicity concentrations range for Cu, Fe, Mn and Zn were 25-40, 400-1000, 400-2000 and 500-1500 mg kg^{-1}, respectively (Romheld & Marschner 1991). The Cu, Fe, Mn and Zn concentrations in plant shoots were in the normal range and well below the phytotoxic limits. In addition, As, Se and Cd concentrations were not detected in any plant tissues obtained from all substrates. Furthermore, Pb concentration was not significantly differed among all plant tissues. Therefore, growth substrates with 25% and 50% of SA can be recommended as the most effective substrates to substitute expensive and less available peat in environmental point of view. Therefore, synthetic aggregates can be recommended as a viable alternative to substitute expensive peat in horticulture.

4.2.3 Other aggregates as a potting media component

Aggregate formed from sewage sludge with inorganic binder also can be used as potting media component for plant production to substitute congenital expensive peat media. These aggeragets can be used with different amounts of peat, coco fiber or oil palm waste to develop potting media. An experiment conducted with sewage sludge aggregates with different percentage of the peat showed that the highest plant height, number of flowers per plant, fresh shoot weight, shoot dry weight, root length, root fresh weight and root dry weight obtained from the treatment having sewage sludge aggregates and peat at 40% and 60% increased by 13.69%, 23.53%, 41.46%, 58.95%, 2.43%, 39.09% and 21.68%, respectively compared to peat control. Moreover, sewage sludge aggregates addition increased the N, P, Ca and Mg contents in the plant tissues. The aggregates application did not significantly increase the Cu, Cd, Cr and Pb in the plant tissues but accumulation of Zn was increased significantly. Aggregate addition did not pose any phytotoxicity in French marigold plants. Therefore, sewage sludge aggregates can be suggested as a viable potting media component to substitute widely using expensive peat in horticulture. Synthetic red soil aggeragets with different types of compost (sewage sludge sugarcane trash compost, cattle manure compost, chicken manure compost etc.) can also be used as a potting media for plant production. Increased biomass production of lettuce was reported from the sewage sludge sugarcane trash compost and synthetic red soil aggregate assayed media compared to peat media. Moreover, due to physical and chemical characteristics of the media developed by sewage sludge sugarcane trash compost and synthetic red soil aggregate based media can be considered as valuable partial peat substitutes for lettuce, especially at the rates of 40% of sewage sludge sugarcane trash compost, 20% of red soil aggregates and 40% of peat, which gave the maximum growth parameters and the highest biomass yield of the lettuce when compared to peat (Jayasinghe et al., 2009f). In addition, red soil synthetic aggregates and zeolite can be used as an alternative medium for French marigold production which gave higher growth and yield (Jayasinghe et al., 2009g). Therefore, potting media developed by using sewage sludge, sugarcane trash, paper waste and low productive soil can be considered as a suitable method for recycling and reducing the environmental impact of these residues.

5. Conclusions

Synthetic aggregates can be developed from wide range of wastes types and can be utilized in different ways. Different types of aggregates can be developed according to the required utilization method. Moreover, the sizes of the aggregates can be decided with respect to the requirement. Aggregates can be designed according to the availability of the waste materials in the area or the region. Developed aggregates can be utilized as a soil amendment to problematic soils to enhance their challenged physical and chemical characteristics. Coal fly ash aggregates as a soil amendment to acidic soil improve the soil poor properties and subsequently improved the crop growth parameters. In addition, synthetic aggregates effectively can be used as a potting media component to substitute conventional media. Aggregates addition to the soil as an amendment improved the microspores due to microbial activities with the assistance of the fibrous materials in the aggregates, which lead to enhance the soil water holding capacity, porosity and permeability. Synthetic aggregate production from different wastes can be recommended as a potential way of waste management. Waste materials contain considerable contents of trace elements that can be detrimental to human and therefore routine investigations should be undertaken to get confirmed the trace element concentrations are below the permissible levels in the waste materials. Future studies should be undertaken to study the potential utilization of these aggregates for different range of crop varieties.

6. Acknowledgements

Most sincere thanks and appreciation are extended to the Higher Education Ministry Japan (Monbukagakusho) for the financial support given to conduct relevant studies. A very special thank is extended to Mrs. Inoka Dharshani Liyana Arachchi for excellent support given to write this chapter. Thanks are also due to all members of Department of Environmental Science and Technology, Faculty of Agriculture, University of the Ryukyus, Japan for their assistance given through out the research period.

7. References

Abad, M., Martinez-Herrero M. D., Martinez Garcia J & Martinez C. (1993). Evaluacion agronomica de los sustratos de cultivo (Agricultural evaluation of crop media).I Jornadas de sustratos. Actas de Horticultura,11:141-154, SECH.

Abad, M., Noguera P., & Bures. S. (2001). National inventory of organic wastes for use as growing media for ornamental potted plant production: case study in Spain, Biores.Technol.77, 197–200.

Adesodun, J. K., Mbagwu, J. S. C. & Oti, N. (2005). Distribution of carbon, nitrogen and phosphorous in water stable aggregates of an organic waste amended Ultisol in Southern Nigeria. Bioresour. Technol. 96, 509-516.

Adriano, D. C. (2001). Trace Elements in Terrestrial Environment. Biochemistry, Bioavailability, and Risks of Metals. Springer, New York.

Adriano, D. C., Page, A.L, Elseewi, A. A, Chang, A. C & Straughan, I. (1980). Utilization and disposal of fly ash and other coal residues in terrestrial ecosystems: a review. J. Environ. Qual.9, 333–344.

Aitken, R. L., Campbell, D. J., & Bell L. C. (1984). Properties of Australian fly ash relevant to their agronomic utilization. Aus.J.Soil Res.22, 443-453

Allison, F. E. (1973). Soil organic matter and its role in Crop Production. Elsevier Scientific, New York, USA. Ben-HurM, Keren R. (1997). Polymer effects on water infiltration and soil aggregation. Soil Sci. Soc. Am. J. 61, 565-570.

Benito, M., Masaguer, A. Moliner, A. & Antonio, R. D. (2006). Chemical and physical properties of pruning waste compost and their seasonal variability Bioresour Technol. 97, 2071-2076.

Chaney, K., & Swift, R. S. (1984). The influence of organic matter on aggregate stability in some British soils. J.Soil.Sci.1984 (35), 223-230

Chapman, H. P. (1966). Diagnostic Criteria for plants and soils, University of California. Division of Agricultural and Natural Science. Oakland, CA, USA.

Chen, J., & Li, Y.(2006). Coal fly ash an amendment to container substrate for *Spathiphyllum* production.Bioresour Technol.97, 1920-1926

Davison, R. L., S. Natusch, D.F., & Wallace, J.R, Evans, C. A. (1974) Trace elements in fly ash; dependence of concentration on particle size. Environ. Sci.Technol. 8, 1107-1113

De Boodt, M. & Verdonck, O. (1972). The physical properties of the substrates in horticulture, Acta. Hort.26, 37–44

Dexter, A. R. (1988). Advances in characterization of soil structure. Soil Till Res.11, 199-238.

Elena, S. (2004). The contribution of waste management to the reduction of greenhouse gas emissions with applications in the city of Bucharest. Waste Management & Research. 22, 413-426.

EL-Mogazi, Lisk, D. J. & Weinstein,L. H. (1988). A review of physical, chemical and biological properties of fly ash and effects on agricultural ecosystems. Science of the Total Environment. 74, 1-37.

Elseewi A. A., Straughan I. R., & Page A. L. (1980). Sequential cropping of fly ash amended soils: effects on soil chemical properties and yield and elemental composition of plants. Sci. Total Environ. 15, 247–259.

EPA. (2008). Municipal Solid Waste. Basic Information. US Environmental Protection Agency. http://www.epa.gov/msw/facts.htm.

Epstein, E. (2003). Land Application of Sewage Sludge and Biosolids. Lewis Publishers, Boca Raton, USA.

Fischer, G. L., Chang, D. P. Y., & Brummer, M. (1976). Fly ash collected from electrostatic precipitator. Microcrystalline structures and the mystery of the spheres. Science.19, 553-554

Gil, M. V., Calvo, L. F., Blanco, D., & Sanchez, M. E. (2008). Assessing the agronomic and environmental effects of the application of cattle manure compost on soil by multivariate methods. Bioresour. Technol.99, 5763-5772

Green, V. S., Stott, D. E., Norton, L. D., & Graveel, J. G. (2000). Polyacrylamide molecular weight and charge effects on infiltration under simulated rainfall. Soil Sci. Soc. Am. J. 64, 1786-1791.

Hamazaki, T. (1979). Parent materials and soils of Nansei-shoto in Japan, Pedologist. 23, 43–57.

Hodgson, D. R. & Holliday, R. (1966). The agronomic properties of pulverized fuel ash. Chem. Ind. 20, 785–790.

Jala, S. & Goyal, D.(2006). Fly ash as a soil ameliorant for improving crop production—a review. Bioresour Technol. 97, 1136–1147

Jayasinghe, G. Y., Tokashiki, Y., & Kitou, M. (2005). Development of synthetic soil aggregates utilizing coal fly ash and mine clay as waste materials. Tropical Agriculture Research and Extension. 8, 1-12

Jayasinghe, G. Y. & Tokashiki, Y. (2006). Utilization of coal fly ash for the production of artificial aggregates as a crop growth medium with acidic "Kunigami Mahji" soils in Okinawa, Japan. Journal of Solid Waste Technology and Management.32, 1-9.

Jayasinghe, G. Y., Tokashiki, Y., & Kitou, M. (2008). Characteristics of synthetic soil aggregates produced by mixing acidic "Kunigami Mahji" soil with coal fly ash and their utilization as a medium for crop growth, Soil.Sci.plant.Nutr. 54, 264–276.

Jayasinghe, G.Y., Tokashiki, Y., & Kitou, M. (2009a). Evaluation of coal fly ash based synthetic aggregates as a soil ameliorant for the low productive acidic red soil, Water Air Soil pollut. 2009(204), 29-41. DOI 10.1007/sl1270-009-0023-y.

Jayasinghe, G. Y., Tokashiki, Y., Kitou, M. & Kinjo, K. (2009b). Recycling of coal fly ash and paper waste to improve low productive red soil in Okinawa, Japan, Clean Soil Air Water. 37(9), 687-695

Jayasinghe, G. Y., Tokashiki, Y., Kitou, M. & Kinjo, K. (2009c). Coal fly ash based synthetic aggregates as potential alternative container substrate for ornamentals, Plant. Nutr. Soil. Sci. 2009, 172(5), 720- 728. DOI10.1002/jpln.200700296.

Jayasinghe, G.Y., Tokashiki, Y., Kitou, M. & Kinjo, K. (2009d). Effect of synthetic soil aggregates as a soil amendment to enhance properties of problematic grey ("Jahgaru") soils in Okinawa, Japan, Commun. Soil Sci.Plant.Anal. (In press).

Jayasinghe, G. Y., Tokashiki, Y., & Kitou, M. (2009e). Use of soil aggregates as a containerized growth medium component to substitute peat in the ornamental plant production, Archives of Agronomy and Soil Science (In press). DOI: 10.1080/03650340902914024.

Jayasinghe, G. Y., Tokashiki, Y., Liyana Arachchi, I. D., & Arakaki, M. (2009f). Sewage sludge trash based compost and synthetic aggregates as peat substitutes in containerized media for crop production. J.Hazard.Mater 2010 (174), 700-706: DOI; 10.1016/j.jhazmat.2009.09.107

Jayasinghe, G. Y., Tokashiki, Y., Liyana Arachchi, I. D. & Kinjo, K. (2009g). Evaluation of the use of synthetic red soil aggregates and zeolite for ornamental plant production. Journal of plant nutrition. (In press).

Klose, S., Tolle, R., Baucker, E. & Makeschin, F. (2003). Strategic distribution of lignite-derived atmospheric deposits in forest soils of the upper Lusatian region, East Germany. Water, Air, and Soil pollution, 142, 3-25

Kobayashi, T. & Shinagawa, A. (1966). Studies on the soils of Nansei Islands in the subtropical region of Japan. Bull. Fac. Agric. Univ. Kagoshima.16, 11–55.

Lee S. B., Lee, Y. B., & Lee, C. H. (2008). Characteristics of boron accumulation by fly ash application in paddy soil. Bioresour Technol. 99(13):5928–32.

Loneragan, J. F. & Snowball, K. (1969). Calcium Requirements of Plants, Aus.J.of Agric.Res. 20, 465-478.

Louis, G.E. (2004). A historical context of municipal solid waste management in the United States. Waste Management and Research. 22, 306–322.

Lynch, J. M. & Bragg, E. (1985). Microorganisms and soil aggregate stability. Adv. Soil.Sci.198, 133-171.

McCarthy, J. E. (2007). Interstate Shipment of Municipal Solid Waste: 2007 Update.CRS Report for Congress. RL34043. Resources, Science and Industry Division, Washington, DC, USA.

Melosi, M. V. (1981). Garbage in the Cities: Refuse, Reform and the Environment, 1880–1980. Texas A&M University Press, College Station, USA.

Meyers, G. D., McLeod, G. & Anbarci, M. A. (2006). An international waste convention: measures for achieving sustainable development. Waste Management and Research. 24, 505–513.

Molliner, A. M. & Street, J. J. (1982). Effect of fly ash and lime on growth and composition of corn (*Zea mays* L.) on acid sandy soils. Proc. Soil Crop Sci. Soc. 41, 217–220.

Moriguchi, Y. (1999). Recycling and waste management from the viewpoint of material flow accounting. Journal of Material Cycles and Waste Management 1, 2–9.

National Institute of Agro Environmental Sciences. (1996). Classification of cultivated soils in Japan, Third Approximation, National institute of agro-environmental sciences, Tsukuba, Japan.

NRC (2000). Biosolids Applied to Land: Advancing Standards and Practices. National Research Council (NRC), National Academy Press, Washington, DC, USA.2000.

Nyakatawa, E. Z., Ready, K.C. & Sistani, K.R. (2001). Tillage, cover cropping and poultry litter effects on selected soil chemical properties. Soil Til.Res. 58, 69-79.

Okinawa Prefecture Agricultural Experiment Station. (1999). Soils in Okinawa prefecture and measures to improve their quality, Okinawa International Center, Japan international cooperation Agency (JICA), Okinawa, Japan.

Pathan, S. M., Aylmore, L. A. G., & Colmer, T. D. (2003). Properties of several fly ash materials in relation to use as soil amendments. J. Environ.Qual. 32, 687-693.

Peterson, J.C. (1982). Modify your pH perspective. Florists' Rev.169 (34/35), 92-94

Rasp, H. & Koch. E. (1992). Waste product from the paper production and their use in agriculture-An experience report. In. Conference Proc. on Waste material from paper manufacture and its utilization in agriculture one experience. VDLUFA-Kongress, Gottingen, Germany, 14-19, September, pp 345-348.

Raviv, M., Chen, Y. & Inbar, Y. (1986). Peat and peat substitutes as growth media for container-grown plants. In: Chen, Y., Avnimelech, Y. (Eds.), The Role of Organic Matter in Modern Agriculture.Martinus Nijhoff Publishers, Dordrecht, pp. 257–287.

Romheld, V. & Marschner, H. (1991). Functions of micronutrients in plants. In:Mordvedt,J.J., Cox, F.R. Shuman, LM., Welch, RM (Eds.),Micronutrient in Agriculture,SSSA book Series No.4. Madison, WI pp. 297-328.

Sandulescu, E. (2004). The contribution of waste management to the reduction of greenhouse gas emissions with applications in the city of Bucharest. Waste Management & Research. 22 (6), 413-426

Sikka, R. & Kansal, B. D. (1994). Characterization of thermal power plant fly ash for agronomic purposes and to identify pollution hazards. Bioresour. Technol. 50, 269–273.

Simard, R. R., Baziramakenga, R., Yelle, S. & Coulombe, J. (1998). Effects of de-inking paper wastes on soil properties and crop yields, *Can. J. Soil Sci.* 78, 689–697.

Sojka, R. E., Lentz, R. D. & Westermann, D. T. (1998). Water and erosion management with multiple application of polyacrylamide in furrow irrigation. Soil Sci. Soc. Am. J. 62, 1672-1680.

Spiers, T. M. & Fietje, G. (2000). Green waste compost as a component in soil less growing media. Compost Sci.Util, 8, 19-23

Stratton, M. L., A. V. Barker and J. E. Rechcigl. (1995). Compost. In: Rechcigl, J.E. (Ed).Soil amendments and Environmental quality. CRC Press,USA, pp 249-309.

Sugiyanto, Y., Soekodarmodjo, S., Suparnawa, S. H., & Notohadisoewarno, S. (1986). soil Physical properties affecting the rootsdistribution of mature rubber on Red - Yellow Podsolic soil, North Sumatra (Indonesia). Bull.perkaretan Indonesia. 4, 82-88.

Taskin, O., Ali, K. O. & Ekrem L. A. (2002). International Conference on Sustainable Land Use and Management / 2002-Çanakkale, Turkey.

Tejada, M., & Gonzalez, J. L. (2003). Effects of the application of a compost originating from crushed cotton gin residues on wheat yield under dry land conditions. Eur. J.Agron. 19, 357-368.

Tokashiki, Y., Yamada, T., Shimo, M., & Onaga, K. (1994). Physical properties of the surface and the runoff soils of made land in Okinawa Island, Jpn. J. Soil Sci. Plant Nutr. 65, 115–125.

Tripathi, R. D., Vajpayee, Singh, Rai, U. N., Kumar, A., Ali, M. B., Kumar, B. & Yunus, M. (2004). Efficacy of various amendments for amelioration of fly ash toxicity: growth performance and metal composition of *Cassia siamea* Lamk. Chemosphere.54, 1581-1588.

UNEP (1993). Biosolids Management: An Environmentally Sound Approach for Managing Sewage Treatment Plant Sludge - An Introductory Guide for Decision-Makers. United Nations Environment Programme, United Nations University, International Network on Water, Environment and Health, Osaka, Japan.2002.

USEPA (United States Environmental Protection agency).(1993).Standard for the Use or Disposal of Sewage Sludge: Final Rules 40 CFR parts 257,403 and 503.Fed. Reg. 58(32) 9284-9415. USEPA, Washington, DC.

US Environmental Protection Agency (USEPA). (2002). Solid waste management: a local challenge with global impacts. Solid Waste and Emergency Response. EPA530-F-02-026 (5306W) second ed.

US Environmental Protection Agency (USEPA). (2006). Inventory of U.S. Greenhouse Gas Emissions and Sinks.

Van Noordwijk, M., Schonsderbeek, D. & Kooistra, M. J. (1993). Root-soil contact of grown winter wheat.Geoderma. 56, 277-286.

Wong, J. W. C., & Wong, M. H. (1990). Effects of Fly ash on Yields and Elemental Composition of Two Vegetables, *Brassica parachinensis* and *B. chinensis*. Agr. Ecosyst. Environ. 30, 251–264.

Wright, A. L., & Hons, F. M. (2005). Soil Carbon and nitrogen storage in aggregate from different tillage and crop regimes. Soil Sci.Soc. Am.J.69, 141-147

Yan, F., Schubert, S. & Mengel, K. (1992). Effect of low root medium pH on net proton release, root respiration, and root growth of corn and broad bean. Plant Physiol. 99, 415-421.